Teaching Business Sustainability Vol. 2

Cases, Simulations and Experiential Approaches

Teaching Business Sustainability

VOLUME 2: CASES, SIMULATIONS AND EXPERIENTIAL APPROACHES

EDITED BY CHRIS GALEA

Greenleaf
PUBLISHING

2 0 0 7

© 2007 Greenleaf Publishing Ltd

Published by Greenleaf Publishing Limited
Aizlewood's Mill
Nursery Street
Sheffield S3 8GG
UK
www.greenleaf-publishing.com

Printed and bound in Great Britain by
CPI Antony Rowe, Chippenham and Eastbourne
Cover by LaliAbril.com.

British Library Cataloguing in Publication Data:
    A catalogue record for this book is available from the British Library.

    ISBN-13: 978-1-874719-73-1

Contents

Introduction

Chris Galea

St Francis Xavier University, Nova Scotia, Canada

I believe that one of the most fundamental trends occurring in our economies—or, at least, in the world's most developed market economies—is that an ever-increasing part of our wealth is being generated by the so-called knowledge economy as opposed to agrarian, resource-based or industrial production. This shift has been taking place since the start of the industrial revolution; its visible result can be gauged by the ever-increasing growth of our service economy as a percentage of all economic activity. Roughly three-quarters of the wealth generated by most developed-market economies is based on services, and that percentage continues to increase inexorably.

While acknowledging that every generation seems to think that it is living in a special time, I believe that this major shift is indeed real and has profound implications for how we manage our companies. Knowledge workers cannot be bossed around like the low-value-added workers of the past. When an ever-increasing part of a company's asset base walks out of the door every evening (or, alternatively, stays at home because they don't actually have to be 'at work'), managers have come to realise that old-style, '2 × 4'[1] management techniques (i.e. where workers are metaphorically hit on the head to get them to perform) are obsolete. When workers add value by their brains rather than their brawn they need to be treated much more as equals and partners; the key task of managers now is to create an environment where creativity and ingenuity can flourish rather than enforcing rote adherence to standard operating procedures.

The effects of this shift into a knowledge-based economy are starting to affect all aspects of the corporation and how it is managed, albeit more slowly than perhaps desired. (*The Economist* [2006], in a recent survey titled 'The New Organisation: A Survey of the Company', laments that 'The way people work has changed dramatically, but the way their companies are organised lacks far behind.') Nevertheless, change com-

---

1   A '2 × 4' is a standard size of construction timber in North America (it actually measures $1\frac{1}{2}$ inches by $3\frac{1}{2}$ inches—please don't ask why!).

panies must, otherwise they risk joining the fate of so many other mighty leviathans that have fallen on their organisational swords.

Likewise, the teaching of business (and, for that matter, the business of teaching) must adapt to the dramatically different learning needs that such a profound shift demands. This change is especially needed when it comes to teaching business sustainability, for if there is one area of business that requires, *ipso facto,* out-of-the-box, creative thinking it is precisely this one. Business sustainability, by its relative newness (and hence uncertainty), its dependence on interdisciplinary thinking, its need to work with different stakeholders, its non-traditional operating approaches and so on, demands that we train our managers in wholly new ways. Beating our students with the proverbial 2 × 4 rote subject matter is no way to inculcate the kinds of multi-varied, creative skills and approach that such a subject demands.

This need for new and non-traditional teaching approaches is reflected in this collection of unorthodox teaching pedagogies. The underlying philosophy behind them is that deep learning for sustainability needs ultimately to be experiential: that is, learning while doing rather than a passive absorption of facts and figures. While much of the underlying theory of sustainability may be taught using more traditional lecture and reading approaches, the implementation of true business sustainability requires students to experiment—to win and lose—while grappling with the myriad challenges and frustrations posed by sustainability: the same challenges and frustrations, one might add, that companies bent on implementing sustainability face on a daily basis in the outside world in which they operate.

Experiential learning: why it works

Experiential learning—role-plays, case studies, simulations and so on—is effective for a number of different reasons, most of which revolve around the idea of bringing the so-called 'real world' into the classroom. For example, role-plays and simulations 'model some aspect of reality in a safe and time-compressed setting' (Hequet 1995). A key component of a simulation's effectiveness as a learning tool comes directly from how well it recreates the world it is simulating, while lessening the inherent risks (Maital-Shlomo and Morgan 1988). Likewise Foxon (1990) reports that one of the fundamental principles underlying the use of simulations 'is that a commitment to experiential learning [and] training that simulates the real world is a powerful learning tool'.

Of course, experiential teaching methods cannot totally recreate every single aspect of the real world. However, failure in a real-world career setting often has painful negative personal and organisational consequences. Also, high-risk actions are necessarily tempered because of the real potential for negative outcomes. As may be expected, learning under such circumstances is naturally constrained. However, in a simulated experiential setting the risks of failure are low to non-existent; under such circumstances learning from failure becomes non-threatening and hence inherently more effective.

Because experiential learning is a two-way street (or, more often, a multi-way avenue), the learning it generates relates directly to the weaknesses of the participants.

It thus helps them to better understand and perhaps address any key deficiencies identified. This point is emphasised by Keys: 'simulation games used for [management] training . . . are very flexible learning environments allowing different students to learn different things' (Keys 1986).

There are many other reasons why experiential learning has many devoted followers. Many of these pedagogical merits have been documented elsewhere; however, one that is often overlooked is that teaching and learning in an experiential setting is great fun—for both teachers and students alike. And there's nothing like having fun to really generate one's creative juices and to make real learning stick.

What follows from the thinking outlined above are the experiential teaching approaches gathered together in this volume—which is the companion to the collection published in 2004 under the title *Teaching Business Sustainability Volume 1: From Theory to Practice* (Greenleaf Publishing; www.greenleaf-publishing.com/catalogue/tbsus1.htm).

The various teaching approaches fall into four main categories of experiential pedagogy: active-learning teaching approaches, hands-on exercises, case studies and role-play simulations. Each is briefly described in more detail below.

Active-learning teaching exercises

The first contribution in the active-learning category is called 'Getting out there: incorporating site visits and industry assessments in pollution prevention and sustainability education'. The exercise was developed by Kim Fowler of the Pacific Northwest National Laboratory in Richland, Washington, and Jill Engel-Cox of the Battelle Memorial Institute in Arlington, Virginia. In their chapter they present an excellent, hands-on exercise that provides students with a clear connection between the theoretical concepts they are learning and their application in the 'real' world. Rather than trying to simulate the challenges, barriers and incentives of conducting an environmental assessment in a classroom setting, the authors show how to incorporate an environmental assessment of an actual industry or small business directly into the students' training. The assessments can range from a simple site visit followed by in-class discussion to multiple site visits with written analytical reports. While acknowledging the inherent uncertainties of this method, the authors contend that the exercise can be challenging for teacher and students alike. They make a strong case for showing the large learning potential that these hands-on exercises can generate.

The second chapter in this section is titled 'Different planets: belief, denial and courage. The role of emotion in turning learning into action'. It is written by Penny Walker, an independent consultant specialising in helping people in business learn about sustainable development and putting that learning into action.

In her study Walker describes the role of emotional responses to evidence and rational arguments. It is clear that there are people who accept the evidence of our current unsustainability and yet choose not to act on it. She argues that emotional responses affect the extent to which participants engage with, and take action on, sustainability. Her study describes ways to help participants bridge the gap between knowing and doing, acknowledging rather than ignoring how people feel about what they have learned. In trying to understand this, the author has drawn on models of adult learning, group facilitation and responses to terminal illness. In her workshops Walker uses techniques such as incisive questions, facilitated discussion and action planning.

The next teaching approach is by Shirley Eber who focuses on sustainability in business education as it relates to the leisure and tourism industry. Shirley echoes the hope of the 1999 UK Secretary of State for Education and Employment who proposed that 'all [those] involved in the education of today's business students will develop and implement sustainable development education strategies'. The author contends that tourism makes an ideal case study for such a project, for at least two major reasons: the global nature of tourism (a truly global business and also the world's largest industry according to the World Tourism Organisation) and its increasing popularity as a vocational area of study. The author then shows why and how sustainability can and should be integrated into business education, and outlines her case using this industry as an example.

The last teaching approach looks at the challenges, methods and tools in teaching sustainability. In 'Getting it: understanding the science and principles behind sustainability', the authors Darcy Hitchcock and Marsha Willard share with us their experience in the field of teaching business sustainability. They aim to give students who are new to the concept of sustainability 'a grounding in its definition and the underlying scientific principles', thus helping them to 'get it'. The authors contend that learning the science (even to non-scientists) is easy; the real challenge is overcoming misconceptions, prejudice, disinformation and guilt. This approach addresses these specific obstacles to learning. It is then followed in the next section with two hands-on exercises that specifically address the issues raised here.

Hands-on exercises/activities

The first two hands-on exercises have both been developed by Darcy Hitchcock and Marsha Willard; each follows from their teaching approach outlined in the end of the previous section. The first activity is called 'The Sustainability Card Game'. This activity gives learners the necessary grounding in the science of The Natural Step System Conditions and does so in an engaging, interactive and experiential approach. The second activity is 'The Multi-organisation Business Simulation'. The purpose of this exercise is to 'give people an opportunity to apply sustainability and The Natural Step System Conditions in a complex, realistic, but hypothetical situation'. The exercise, when combined with the one just described, achieves this goal admirably.

The other exercise is 'Personalising sustainability: an interactive activity to reinforce the presentation of The Natural Step (TNS)'. This exercise was developed by Joshua Skov of Good Company, Eugene, Oregon, and helps educators to introduce the concepts of sustainability to any audience by addressing both its intellectual and emotional challenges. Clearly, for most students of sustainability there are large information gaps to be filled. But information is not enough to make us feel a personal link between our behaviour, the health of society and the planet. The challenge in presenting the concept of sustainability lies in the complexity of our interdependence with each other and the planet.

Skov presents an activity for helping an audience to feel the relationship between sustainability issues and individual behaviour. Using a straightforward, single-page handout for each person, the exercise leads a group through the mental process of tying the System Conditions of The Natural Step to several commonplace individual behaviours.

The author reports that the activity succeeds for three distinct reasons. First, it gently and effectively removes the attention from the presenter/lecturer and places it on the audience members, individually and collectively. Second, in less than an hour, it provides learning in both individual and group modes. And, third, it requires individuals to internalise the links between TNS System Conditions and everyday activity, by presenting 'findings' (i.e. the information from and insights based on their respective handouts) to each other in a non-threatening, small-group setting. This learning-by-teaching leads the participant to internalise that knowledge and take the other participants' individual teachings more seriously.

Case studies

This section presents a number of case studies, each aimed at teaching various aspects of sustainability.

The first case, 'Easter Island: a case study in non-sustainability', is written by Canada's well-known demographer David Foot. It focuses on the recently documented history of Easter Island, that enigmatic rock 3,000 km off the coast of Chile. The case provides students with 'a simple, easy-to-understand historical case study of an isolated society growing and developing but not practising sustainability through ignorance, neglect, self-interest or simply bad luck'. At first the case appears deceptively simple— the study of yet another ancient people falling prey to environmental overreach. And yet, with further in-depth thinking, the power of the case emerges as students absorb the important lessons of Easter Island for the similarly isolated inhabitants of planet Earth today. By drawing such a clear parallel the case also points to the ominous consequences of not practising sustainable behaviour for both individuals and society. It makes a good start to any course on sustainability.

The next case is 'Suncor Energy: a comprehensive approach to sustainability'. In this chapter the authors, Peter Stanwick and Sarah Stanwick of Auburn University in Alabama, look at the challenges faced by a company that is trying to be a leading-edge force for sustainability in an industry that is synonymous with pollution and environmental degradation. While some would reject out of hand the notion that an energy company can ever be sustainable, the Suncor case shows how it is possible for a firm to move towards a trajectory of sustainability.

In a different vein, Kumba Jallow of De Montfort University in the UK looks at 'Perspectives on accounting and society. Teaching accountants corporate social responsibility (CSR): a case study'. She believes that 'changes in the business world should be reflected in developments in undergraduate university courses'. She has set out to do so in the field of accounting as it relates to sustainable development and corporate social responsibility. In this case she outlines the challenges and successes she has encountered over the years of teaching this subject.

The next case study is—though I declare an interest—one of the more 'fun' cases in the book. It outlines the development of 'The Anti-Junk Mail Kit': a product/service designed to help people eliminate unwanted and unsolicited junk mail from their lives. It recounts the entrepreneurial opportunity seized upon by two enthusiastic business graduates and their deliberations as they develop the company. The case gives students the opportunity to analyse the various challenges and options faced by the entrepreneurs and to make decisions as to how they would go about developing the company.

The last case is an in-depth decision exercise written by Monika Winn of the University of Victoria and Charlene Zietsma from the University of Western Ontario. It presents the student with the stark choices facing MacMillan Bloedel (MB), then Canada's largest and most visible wood-products company. The student is asked to determine what MB should do in the face of terrible financial pressures and an unprecedented and increasingly successful campaign by environmentalists to encourage customers to boycott MB's products. The case presents a complex decision problem in which decision-makers must take into account the needs of multiple stakeholders with conflicting interests. This presents an excellent opportunity for students to practise role-playing skills ranging from stakeholder management to conflict resolution techniques.

Role-play simulations

The last section presents three role-plays that incorporate many of the approaches outlined above.

The first role-play presented here is called 'Sustainable games people play: teaching sustainability skills with the aid of the role-play 'NordWestPower'. It is co-written by Anke Truscheit and Christoph Otte of the University of Oldenburg, Germany.

NordWestPower is a behaviour-oriented role-play that trains managers to evaluate the social and ecological risks of their field. The authors contend that lectures, seminars and other conventional teaching styles alone are not adequate for teaching these necessary skills; they need to be supplemented with experiential approaches. Using the role-play NordWestPower they deliver an overview of the ideas and structure behind role-plays and the possibilities they open up for teaching sustainability skills.

As may be expected, the role-play NordWestPower deals with the issues faced by a European energy provider. However, the structure of the role-play is very flexible and is easily adapted to other industries or environments. The authors lay out the various stages of the simulation and give clear guidelines to educators on how to use the role-play in their own classrooms and what the expected learning outcomes should be.

The second simulation exercise, 'Using experiential simulation to teach sustainability', is written by two management consultants and trainers, Susan Svoboda and John Whalen. Their chapter focuses on a simulation they developed called the 'Transformation exercise'. Transformation is a reality-based, team-building simulation that helps participants understand how to translate the concepts of sustainability into tangible action. The authors offer a practical guide for using experiential simulations to teach sustainability in a business context. They draw on their experience in developing and using the exercise with hundreds of groups in business schools, corporations, government organisations and non-profit organisations. They describe the benefits of using simulation in building understanding of sustainability, the structure and process of the simulation, discuss the typical lessons learned, and provide a checklist of the characteristics of an effective sustainability simulation.

The last role-play/simulation is one developed by David Annandale and Angus Morrison-Saunders, both of Murdoch University in Australia. It is entitled 'Teaching process sustainability: a role-playing case focused on finding new solutions to a waste-water management problem'. The key learning objective of this simulation is to move learners beyond focusing solely on 'content' issues (e.g. reduce production of wastes or energy conservation) and to think about the *processes* by which we move towards sustainable outcomes. These processes themselves also need to be sustainable. The authors examine this link between sustainable outcomes and process sustainability by way of a simulation case study. The 'outcome focus' of the role-play is 'the attempt to determine a new, and more sustainable, waste-water treatment option for a regional town in a particularly beautiful part of southern Western Australia'. The interactive, role-playing approach investigates the problem by assigning stakeholder roles to students. The case allows students to 'reassess the way government bureaucrats made their original decision', and presents them with the information to search for a better solution. It challenges students to think about whether it is actually possible to 'find a sustainable solution to a problem, without addressing the sustainability of the decision-making process'.

Conclusion

I hope that the approaches and exercises will whet your teaching appetite for incorporating more experiential pedagogy into your curriculum. Yes, the learning outcomes and teaching approach will not be as predictable and structured as you may be used to. Yes, you will run into novel and perplexing learning situations. Yes, you do have to give up some control over the learning environment. These are all valid fears arising from incorporating experiential learning into the curriculum. However, adopting an experiential approach leads to a much greater probability of creating a learning environment where a whole series of proverbial shoes may drop as the students themselves take control over their own learning. This is the way it should be for, ultimately, true teaching is far less about passing on knowledge and far more about helping students learn for themselves.

References

Economist (2006) 'The New Organisation: A Survey of the Company', *The Economist*, 21 January 2006.

Foxon, M. (1990) 'Using Simulations to Enhance Technical Training', *Journal of European Industrial Training* 14.4: 17-20.

Hequet, M. (1995) 'Games that Teach', *Training* 32.7: 53-58.

Keys, B. (1986) 'Improving Management Development through Simulation Gaming', *Journal of Management Development* 5.2: 41-50.

Maital-Shlomo, M., and K. Morgan (1988) 'Playing at Management', *Across the Board* 25.4: 54-62.

Part I
Active-learning
teaching exercises

1
Getting out there
INCORPORATING SITE VISITS AND INDUSTRY ASSESSMENTS IN POLLUTION PREVENTION AND SUSTAINABILITY EDUCATION

Kim Fowler

Pacific Northwest National Laboratory, USA

Jill Engel-Cox

Battelle Memorial Institute, USA

Environmental assessments (pollution prevention opportunity assessments or sustainability assessments) are used by industry to optimise their operations. These assessments involve characterising the process, identifying potential alternatives, evaluating the alternatives and making recommendations. The typical benefits associated with performing assessments and implementing the identified opportunities include cost avoidance, material use reduction, waste minimisation, risk reduction and good public relations. We have used environmental assessments at community organisations and businesses as a tool to take the pollution prevention, sustainable design and cleaner production concepts learned in the classroom into a practical setting.

In our experience, students in both business training and academic courses value a clear connection between the theoretical concepts they are learning and their application in the 'real' world. Science and engineering play a significant and growing role in people's lives yet there is a worldwide trend of declining enrolment in these topics (ICS and WFEO 2002). Enquiry-based, hands-on science education creates a personal experience for students to associate with the content provided in the classroom. Immediate, practical application of the theories learned in the traditional academic setting connects theory and practice, becoming the basis for critical, reflective thinking and action (Lemke 1992).

Pollution prevention, sustainable design and cleaner production demonstrate their main value through industrial application; thus, it is even more important for students to get direct personal experience in these subjects. For example, the value of a pollution prevention activity (such as replacing hazardous chemicals with non-toxic substitutes) changes with a specific business application because of a number of variables (e.g. scale of the technology, cost of labour, local community and regulatory drivers, availability of the technology, etc.). This means that the 'correct' or best answer depends on how well the student identifies the variables, gathers data related to those variables and/or works from an incomplete set of data. In our experience, learning how to identify and analyse those variables is an essential skill for implementing pollution prevention/sustainable design, and is easier for students to grasp when they are working with a specific application.

Additionally, although the basic technical concepts behind pollution prevention and sustainable design have remained constant and are relatively easy to explain to students, technology, product availability, product quality, and the application of pollution prevention and sustainable design options have changed significantly over the last decade. Providing descriptions of the technologies and their applications, then teaching the students how to find and keep up with the latest innovations and how to evaluate new technologies are key topics in any pollution prevention or sustainable design/development course. The students learn the mechanics of identifying, analysing and making pollution prevention recommendations in the classroom. Our experience is that, until they have the practical experience applying pollution prevention and sustainable design recommendations to a specific, real-life situation, they do not understand the complexity of the process.

For example, we have offered pollution prevention overview training to field staff at a large research and development organisation both *with* and *without* the hands-on activities. When offering instruction *without* the hands-on application, the training provided the basics of pollution prevention, sustainable design and cleaner production in a half-day class and used verbal examples and anecdotal stories specific to that organisation. The stories were offered by peers who had first-hand experience with performing environmental assessments plus examples from the course presenters. At the end of the course, the students tended to focus on treatment and recycling opportunities that required minimal interaction with others. The individuals that were trained *with* direct experience of performing an environmental assessment were more aggressive at identifying opportunities for pollution prevention and sustainable development for their clients and co-workers and have received considerable recognition for those efforts. One of those participants told us:

> I thought I understood pollution prevention. Then I did an assessment and realised that it wasn't as simple as selecting a solution from a list of possible technologies. Now I see opportunities in the laboratories that I didn't see before.

All types of organisations, including manufacturing processes, laboratory operations, small businesses, offices and schools, can use this assessment methodology. However, employing the methodology takes environmental/pollution prevention knowledge, assessment methodology experience and time. Owing to the level of expertise needed, smaller organisations are typically not able to perform these assessments without some form of assistance. To perform an assessment you need the following:

- Pollution prevention, sustainable design and cleaner production expertise
- Time to evaluate the opportunities that are identified
- Knowledge of the organisation
- Desire to decrease cost and/or positively impact the environment
- Willingness and ability to effect change when recommendations meet the organisation's needs

Small organisations tend to know their business very well and, if they have volunteered for this type of assessment, they are interested in improving their processes especially if it will reduce their operating costs. However, they may not have the staff available with the expertise and/or the time to focus on pollution prevention activities.

The needs of smaller organisations for environmental technical assistance and for students to directly experience pollution prevention can both be met through well-designed training sessions or university-level coursework. For example, having under-graduate and/or graduate students perform pollution prevention or sustainability assessments for local businesses and schools within the structure of a course provides a quality learning experience for the students and useful information for the business.

There are also indirect benefits to a university in having the students conduct these assessments. By providing pollution prevention and sustainable design technical expertise to the community, they are offering a service that may not be available to local organisations. In a resource-limited society, these assessments can be used to address specific concerns of the community such as water conservation or hazardous waste minimisation. This activity can also be interpreted as 'free' promotion of the university for the education programmes it offers. The students and instructors demonstrate the value of continued education to their everyday activities, which can result in increased numbers of students at the university and a positive image in the community.

The activity

Over the past seven years, we have taught a combined undergraduate/graduate-level course on the principles of pollution prevention that has a community-based pollution prevention opportunity assessment as its core project. In preparation for each semester, we engage local businesses and organisations through contacts with the city government and other community entities. Each organisation agrees to meet with the students two or three times and provide them with information about their operations; this is described in a short release form that they sign. In return, they receive a final pollution prevention assessment report prepared by the students. Once the semester begins, students sign up in teams of 2–4 for the assessment of their choice. Lectures cover the basics of pollution prevention and the steps in the assessment/audit process. The assessment is broken into five steps—background, process mapping, opportunity assessment, cost–benefit comparison and recommendations—which we require the students to hand in step by step over the last ten weeks of the semester. At the end of the semester, the teams present their results to the class (and the community organisation, if it attends) and the final report is given to the participating organisation. While we have experienced the occasional challenge of miscommunication between student and business owner, students have been generally positive about their involvement in the community and in gaining an understanding of the priorities of running a small business. The community organisations receiving the assessment appreciate the straightforward cost and waste reduction approach and the involvement with a local university. We have conducted assessments at a diverse set of locations including a restaurant, beauty salon, funeral home, sanitary waste collection company, a school district, landscaping firm and a child care facility design.

The assessment approach is an application of the sustainability and pollution prevention technical information that the students are first taught during class lectures. Students are provided with overall definitions, commonly used techniques and tools, resources for new technologies and approaches, and a methodology for performing the assessment. The methodology is based on a step-by-step process that includes the following (see Fig. 1.1):

1. Obtain organisational commitment to performing an assessment

2. Gather data about key waste generating and material use activities by reviewing existing documentation, material usage, and past and anticipated waste streams

3. Identify potential pollution prevention opportunities

4. Research and analyse pollution prevention opportunities for potential waste reduction, cost avoidance and return on investment

5. Make recommendations for pollution prevention implementation projects based on the waste and cost analysis

6. Summarise the findings

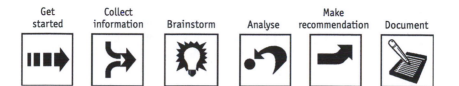

FIGURE 1.1 Steps in a pollution prevention design assessment

Source: Engel-Cox and Fowler 1999; graphic reproduced with permission

The steps that the students take, within the context of the steps listed above, include:

1. Researching pollution prevention techniques already implemented at other organisations that may be similar to the organisation the student will assess

2. Contacting the business to schedule a walkthrough of the process that will be evaluated

3. Preparing a set of questions for the site visit that will be used to elicit potential opportunities and suggestions for improvements from the participants

4. Performing a site visit

5. Reviewing the results of the site visit to identify the top 5–10 opportunities to evaluate

6. Contacting the company to confirm that the items selected would add value to their operations

7. Researching and analysing the opportunities selected, including both technical feasibility and return-on-investment cost assessment

8. Documenting the results in a traditional assessment format (provided for the students)

9. Presenting the results, in both a written report and an oral presentation, to the class and the organisation/site participants

The result of the completed assessment is that the students have learned how to apply the technical knowledge they were taught in the class and the organisations have an assessment they can use to consider implementation investments.

In addition to using this technique for graduate and undergraduate courses in sustainability and pollution prevention, versions of the above have been used for:

- Week-long cleaner production general training which included a half-day site visit followed by in-class brainstorming and role-play

- Two-day assessment training with multidisciplinary teams from the same organisation working on their own industrial processes

A week-long training course can involve a site visit, brainstorming and a rough evaluation of opportunities, allowing the trainees to learn by experience and providing an industry with an initial list of opportunities. When the students are employees of the company they are assessing, the course itself becomes the first step of the assessment process. This technique has been very effective in teaching professionals how to incorporate pollution prevention, sustainable design and cleaner production into their work practices.

For example, we have taught a variety of two- to four-day pollution prevention training courses designed for engineering and environmental professionals. At a government-owned industrial site, a two-day course covered the basics of pollution prevention and assessments in the first morning. The first afternoon included a site tour of the industrial process followed by a brainstorming session. The second day covered prioritisation, estimating waste reductions, cost–benefit analysis and recommendations.

Short discussions of each subject were presented followed by interactive facilitation and activities (e.g. the site tour, brainstorming and group breakout sessions for the opportunity evaluations). Since the students were all from the facility being assessed, they saw the value in the process directly and, in some cases, opportunities that a participant had been considering for some time were given enough analysis (cost and waste reduction) so that very little additional work was required after the course to fund implementation.

A more challenging version of this course was the four-day course taught internationally, with a site visit to a company not associated with any of the students who were from a variety of organisations themselves. With less opportunity for advanced co-ordination, the facilitator for this situation needs to be able to adjust the course and activities quickly depending on the results of the site visit, which on occasion provide much less information or very different information from that expected. In general, the student evaluations emphasise the value in 'getting out of the classroom' and seeing a business in action.

Evaluation of learning

Typically, in the university setting, the assessment constitutes approximately one-third to one-half of the effort and grade for each student, the remainder being exams and homework. The evaluation of the project includes:

- Weekly verbal progress reports (e.g. updates on organisation interactions, selected opportunities, challenges, etc.)

- Portions of the draft report handed in for comment during the course (which ensures the students contact the organisation promptly and allow sufficient time for opportunity assessments)

- Final report

- Presentation to class and organisation

- Group participation

In addition to a grade for the project, the students benefit from this project by learning how to contact and work professionally with an organisation, how organisations operate on a day-to-day basis, and what the priorities are for different types of organisations. With this being taught via example rather than lecture, they are not only learning how to apply the theoretical knowledge, but are also able to contribute to the community. The skills they learn are more than environmental assessments, but also how to adapt ideas, fill information gaps, provide reasonable recommendations, consider the needs of the organisation and create a professional presentation to a client. We often get feedback from students about how they applied pollution prevention to their work or home situations and how their observations have changed. The businesses involved are often implementing the recommendations while the class is ongoing and sometimes the students get involved in helping with implementation. For example, one student helped a local restaurant install insulation. Another helped arrange for cardboard recycling at a local business. The school district implemented all the recommended opportunities before the final presentation was given.

Challenges

Working outside the controlled environment of the classroom has its challenges and requires flexibility in the course outline and adaptability to respond to changes that arise. Adaptability is required from both the students and the course facilitator/ teacher. Table 1.1 details some of the challenges that may be encountered and potential solutions.

Summary

The fields of pollution prevention, sustainable design and cleaner production use environmental assessments to identify opportunities for optimising industry operations. Guiding students through the process of conducting environmental assessments for community organisations is the approach we use to teach students how to apply the techniques and technologies they have learned during a two- to three-month university level course or as part of shorter two-day or one-week training sessions. The assessment approach includes the tasks of gathering background information, mapping the ongoing and anticipated processes, identifying opportunities, evaluating opportunities and making recommendations. The community organisations are selected through contacts with local government and community entities. These organisations commit to providing the necessary information to the students during the time-frame of the course. The students are assessed on their interaction with the organisation, a written report, a class presentation and their level of participation in the group project.

Although performing environmental assessments for real organisations is likely to be more effort for both teacher and students than a traditional lecture-based course, we believe the experiential learning activity is worth the investment for the students, community organisation and the university. The students gain an appreciation of the challenges of working with a company or organisation and the satisfaction of contributing to their community. The organisation that receives the assessment is given information it can use to make educated investment decisions for implementing sustainable practices, helping it to become a better neighbour. And the university benefits through raising awareness of university programmes and contributing environmental assistance to the community.

Challenge	Response
Part-way through the assessment an organisation becomes reluctant to provide the amount of information needed	Make assumptions and document them. Base those assumptions on other sources (e.g. visit similar organisations, obtain case-study information from the internet, simulate the process, etc.) so that the information presented at the end of the assessment has a solid technical basis
The organisation appears to be too 'green' to need any new suggestions	Document the successes that the organisation has already accomplished and look for innovative technologies and solutions that are not as likely to be part of their current organisation. Very rarely is there *nothing* that can be suggested by creative students, although the more innovative suggestions may not be cost-effective. Documenting the possibilities may have value in the future
During the short course, the site tour is cancelled at the last minute	Have enough information prior to the training to simulate the site tour. Alternatively, have a detailed case study to replace the real site visit or conduct an assessment of the building the class is located in. With a short course, having a significant amount of advanced information about a company and preparation of alternative plans are essential
A student on a team does not contribute to the team effort	Make participation a key part of the grade and have team members provide information on their own and their teammates' contributions. By having assignments turned in over multiple weeks and by meeting with the teams, mid-course corrections on product quality and participation can be made as needed
The organisation is not complying with environmental regulations	Prepare and sign an agreement with the organisations prior to performing the assessment stating each group's responsibility and stand by the commitments made in that agreement if regulatory findings are made

TABLE 1.1 Challenges and solutions

References

Engel-Cox, J., and K. Fowler (1999) *Pollution Prevention Opportunity Assessments for Research & Development Laboratories* (Columbus, OH: Battelle Press).

ICS (International Council for Science) and WFEO (World Federation of Engineering Organisations) (2002) *Role and Contributions of the Scientific and Technological Community to Sustainable Development* (E/CN.17/2002/Pc.2/6.Add.8; Johannesburg: World Summit on Sustainable Development, Economic and Social Council, United Nations).

Lemke, J.L. (1992) 'The Missing Context in Science Education: Science', AERA Symposium, Atlanta, Georgia, academic.brooklyn.cuny.edu/education/jlemke/papers/gap-sci.htm, accessed 3 July 2006.

2

Different planets: belief, denial and courage

THE ROLE OF EMOTION IN TURNING LEARNING INTO ACTION*

Penny Walker

Independent Consultant, UK

Learning for change

> **Knowing without doing is not knowing (Chinese proverb).**

Clients bring in someone like me to 'teach sustainability' to people in their organisations because they want to catalyse action. They want to avoid risks (PR disasters, future legal obligations, unnecessary costs), spot opportunities (innovations, market shifts) and understand their stakeholders. Some of them are even explicit that they want to 'save the planet', or at least play a part in doing so.[1]

---

*   *Disclaimer:* The views expressed in this chapter are those of the author alone, and do not necessarily reflect the views of the organisations mentioned.

1   Organisations I have worked with include a leading UK grocery retailer, Carillion Building (a UK construction group) and Interface Europe (part of the Interface floor coverings company). I work both alone, and as a member of a consulting or training team: for example, with The Natural Step UK and Cambridge University's Programme for Industry. I find that clients bring in people like me for one of three primary reasons: strong personal commitment in the most senior parts of the organisation (e.g. Interface); a crisis in the organisation which has woken it up to the business benefits of understanding sustainable development better (e.g. a public relations crisis—Carillion Building is part of what was the Tarmac group, which built the road through Twyford Down, a *cause célèbre* for anti-roads protestors in the UK); or as a 'lateral leadership' approach—in the guise of 'ordinary' environmental awareness training, individuals in the organisation are hoping to catalyse deeper understanding and change.

The learning interventions in which I play a part are not academic exercises. They will have failed if the new knowledge and understanding that participants gain does not lead to a change in their behaviour and in the organisation. In this case study, I describe what I have learned about how to do this more successfully.

The use of evidence, concepts and the intellect

Much of my work is built around making the business case for paying more attention to sustainable development concerns.

The business case is built up in a variety of ways:

- Conceptual
- Theoretical
- Evidential
- Anecdotal

Conceptual

At a **conceptual** level, the metaphor of the Funnel, developed as part of The Natural Step Framework,[2] helps people understand the way in which current trends in environmental capacity and human demand are colliding, and how that reduces a society's room for manoeuvre. It also illustrates the importance of being aware of these trends and planning strategically to avoid decisions that cause you to 'hit the walls' of the funnel (see Fig. 2.1).

There are various ways in which businesses might hit the walls:

- Natural limits: for example, a business dependent on fish finds fish stocks collapsing

- Legislative limits: for example, a business manufacturing a chemical finds it banned

- Public acceptance limits: for example, a business finds people refuse to buy its GM (genetically modified) food, or boycott it because of its labour practices

2   The Natural Step (TNS) Framework was developed by Karl-Henrik Robèrt and there are Natural Step organisations in nine countries around the world. The TNS Framework consists of the Funnel, the Four System Conditions, which describe the conditions for sustainability, and the ABCD process (awareness of the conditions for sustainability, understanding where you are now, envisioning a sustainable future and backcasting from this to determine action to be taken now). For more information about the Framework and The Natural Step's work, see www.naturalstep.org. The Natural Step is an international charity, and in the UK it is part of Forum for the Future.

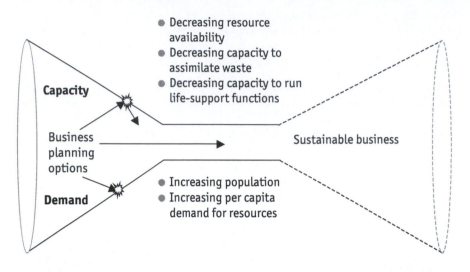

FIGURE 2.1 The funnel

Source: adapted from Walker and Martin 2000

By understanding the Natural Cycle and the four System Conditions that are derived from this (two other key parts of the framework), a business can predict what it should be doing to avoid hitting the walls, and thus continue to be successful.

In theory

The **'in theory'** business case is apparent in various lists of headings, each summarising the ways a business can benefit from improving its sustainability understanding: for example, Forum for the Future's Hierarchy of Business Benefits shown in Table 2.1.[3]

The evidence

Evidence for improved business performance comes in two forms:

- Correlations between good environmental and social performance, and good financial or business performance (such as the Dow Jones Sustainability Index outperforming the 'ordinary' Dow Jones World Index)[4]

- Systematic research, such as SustainAbility's Buried Treasure matrix[5]

---

3 Forum for the Future is a UK charity which works in partnership with businesses, local authorities, academic institutions and NGOs (non-governmental organisations) to 'accelerate the building of a sustainable way of life, taking a positive, solutions-oriented approach'; www.forumforthefuture. org.uk.

4 For the latest correlations, see www.sustainability-index.com/htmle/news/monthlyupdates.html.

5 Buried Treasure examines the links between business success and sustainable development; www.sustainability.com/business-case.

Eco-efficiency	1. Reduced costs
	2. Costs avoided (design for environment, eco-innovation)
	3. Optimal investment strategies
Quality management	4. Better risk management
	5. Greater responsiveness in volatile markets
	6. Staff motivation/commitment
	7. Enhanced intellectual capital
Licence to operate	8. Reduced costs of compliance/planning permits/licences
	9. Enhanced reputation with all key stakeholders
	10. Influence with regulator/government, etc.
Market advantage	11. Stronger brands
	12. Customer preference/loyalty
	13. Lower costs of capital
	14. New products/processes/services
	15. Attracting the right talent
Sustainable profits	16. Option creation
	17. New business/increased market share
	18. Enhanced shareholder value

TABLE 2.1 Hierarchy of business benefits

Source: Cambridge Programme for Industry/Forum for the Future 2001

Anecdotes

Anecdotes of the benefits of 'good' corporate behaviour and the penalties for 'bad' corporate behaviour abound, and help make the theory real.

Intellectual acceptance

Although some people do dispute the evidence and arguments, this is rare. My experience is that, at this stage, most learners accept the broad thrust of the business case arguments (there may be some discussion about points of detail). The debate is around time-scales and their implications for leadership. How soon will the system collapse? When will the business case become real for my industry/organisation? What will our competitors/customers be doing?

Emotional responses

So, at this stage in a programme, there is a group of people who have learned about, and know, the principles of sustainable development. They understand the planetary and business case for change. They do not seriously dispute the evidence of current unsustainability.

How do they feel about what they know?

Broadly, people either feel something, or nothing.

The ones who **feel something** are either energised (they feel courage, curiosity, excitement, motivation) or engaged but disempowered (they feel grief, sadness, fear, impotent anger).

The ones who **feel nothing** seem to be somehow in denial; they have apparently accepted the evidence and the arguments, but they do not *believe* them.

There is a mismatch between intellectual acceptance, emotional response and behaviour change—it's as if people's brains and guts are on different planets.

This phenomenon of denial has been observed and commented on before in relation to sustainability: George Marshall (2001), looking at climate change, drew on the work of Stanley Cohen (who looked at responses to human rights abuses) and identified two key psychological processes:

● When the problem is too awful to accept; he cites Primo Levi: 'Things whose existence is not morally possible cannot exist'

● The 'passive bystander' effect: when many people could act, individuals wait for someone else to act first, 'and subsume their personal responsibility in the collective responsibility of the group'

I have experienced people in groups spontaneously articulating these phenomena. In addition, I have noticed people asking for permission to not believe, and for permission to take on the role of a powerless victim (see Table 2.2).

Passive bystander effect	'[Our company] is avoiding having this conversation—we haven't put it on the agenda—we know individually that it's right, but we are avoiding discussing it.' *Business unit managing director*
Moral impossibility	'If I were convinced that what we are doing now really is unsustainable, I'd think differently.' *Group finance director*
Permission not to believe	'Is it too late?' *Numerous participants, seeking reassurance*
Permission not to be responsible	'They [America, business, the markets, corrupt governments in developing countries] will never allow things to change.' *Numerous participants, blaming others and avoiding responsibility*

TABLE 2.2 Denial in action

As a facilitator how can I respond?[6]

In the early years of my practice, I usually responded with more argument and information, and with somewhat trite reassurances: 'if we all work together then I'm sure we can do it'.

As I practised more and learned more, I realised that these participants weren't speaking from a rational 'place', but from an emotional one.

I realised that these emotional responses are going to be there whether we intend to induce them or not. We don't have a choice about that. What we do have a choice about is whether we take account of this in our practice. Having noticed the different emotional responses, and begun to speculate about their importance, how has my practice changed?

First, I have begun to research existing theories on emotion and change, and learn about how other 'change facilitators' (consultants, trainers, facilitators) see the role of emotion.

Second, I have introduced techniques into my practice, designed to enable people to recognise their own and others' emotional responses, and make them explicit.

I expect to carry on learning more about facilitating for sustainable development, and I do not pretend to have definitive answers. But I have learned a lot during this time, and hope that by sharing some of my learning I can help others to reflect on their own practice.

The theories and insights that have helped me to understand

Many theories and insights have attracted my attention. These seem to be the most important ones.

Kübler Ross's five stages of grief (Kübler Ross 1975)

Developed from work with the terminally ill, Elisabeth Kübler Ross's five stages of grief have been taken up by others as a model of change (both organisational and personal). The five stages are:

- Denial

- Anger

- Bargaining

---

6   You may think of yourself as a trainer, teacher or consultant. I use the term 'facilitator', as the closest to my preferred way of working: enabling discussions that help the group decide what to do with what they've learned, while also offering expert input.

- Depression
- Acceptance

In the face of evidence about calamity, there is a predictable set of emotional responses.

Insights from group facilitation

Exploring this further led me to the substantial body of theory and guidance (see, for example, the work of John Heron [1999]) distilled from experience of facilitating groups.[7] A common theme is the importance of working with the whole system—including the feelings of the participants and the facilitator. 'Feelings are important and need to be acknowledged. They are not rational. Group members need to learn to *have feelings, rather than be had by them*' (emphasis added; Hunter *et al.* 1996: 7).

Observations from other facilitators

As part of my exploration, I have run short sessions on sustainable development for people who are professional trainers and facilitators, but have no professional experience of environmental or social justice issues. After running a half-hour session based around imagining an evacuation of the Earth (designed to bring to conscious awareness people's intuitive understanding of the life-support functions that the Earth provides), I asked the facilitators for their professional feedback. The strength of the emotion generated was a common observation, as was the nature of the emotion—pessimistic (anxiety, fear, guilt) or optimistic (courage, excitement and enthusiasm for pioneering adventure). These personal reflections have an echo in Theodore Roszak's observation: 'But prudence is such a lackluster virtue' (Roszak 1992).

How motivating (and sustainable) are the pessimistic emotions? Can *fear* be the spur? Or are the optimistic emotions the most important in catalysing change?

Transfer of learning

The well-known gap between what people learn on a course and what they do 'back in the office' continues to frustrate managers and facilitators. This is not an issue only when the subject matter is life-changing planetary crisis; it applies equally to the introduction of new internal phone systems or customer service attitudes. There are many strategies that help the successful transfer of learning, and paying attention to emotions is just one of the many helpful things the facilitator can do. Table 2.3 shows some of the critical factors in the transfer of learning, with my own commentary on how they apply to sustainability.

There are features of sustainability as a subject to be learned, and as a set of implied behaviour changes, which heighten the importance of the critical factors:

---

7   Groups with varied purposes: training/learning, therapy, organisational change, consultation/focus groups, or team building.

Critical factor	Description	How the factor manifests in sustainability learning
Reinforcement 'on the job'	The presence or absence of positive feedback from peers and supervisors, when the new knowledge is used the workplace	Many of the people I work with may experience 'lip service' rather than commitment from colleagues and managers
Interference in the workplace	Interference may come from time pressure, those in authority, poor or conflicting work processes, inadequate technology	Add to this cost pressures, low levels of understanding from colleagues, suppliers and customers, and inertia
Organisational culture	The skills or knowledge learned may not sit easily with the predominant culture of the workplace, which may indeed not value learning and change per se	Society's widespread denial of planetary crisis is likely to be present in the workplace
Learner's perception of the need to change	Without a recognition of the need to change, there is no motivation and little readiness to learn	For sustainability, an intellectual acceptance of the need for change may not be a sufficient motivator. Where there is a strong context that favours change, then personal motivation is less important. When the context doesn't favour change—in the absence of strong leadership, legislative push and customer pull—then personal motivation (i.e. emotional engagement) is pre-eminent in determining whether change occurs
Learning objectives that are relevant to the learner	Learning objectives that are 'so specific that they sound odd', and which do not translate into tasks, make learning transfer hard	For sustainability, it is not always clear *what* specific changes in behaviour are required. A sophisticated understanding of the problems can lead to dissatisfaction with superficial behaviour changes. So much behaviour by so many people must change—this uncertainty in itself can generate strong emotions
Rewards	These may be tangible (prizes, bonuses) or intangible (approval of peers and supervisors)	A very few organisations have built sustainability performance into their performance appraisal or rewards and bonus systems. In the absence of this, informal approval (if you like, an emotional benefit) becomes more important

TABLE 2.3 **Sustainability and the transfer of learning**

Source: based on Galbraith 1990 and Taylor 1997

- Dominant trends and drivers reinforce the status quo (almost by definition—otherwise we wouldn't need to teach it)

- Society colludes in the general phenomenon of denial and buck-passing

- The enormity of the problem means that an accurate understanding is likely to be accompanied by strong emotions

- The changes in behaviour implied are at the same time *large* and difficult for the individual, and *insignificant* for the problem if only that individual acts

How has this affected my work?

Theories and insights from other people have helped me reflect on my own work.

Knowing about **Kübler Ross's model**, and that the emotions being felt by group members are common (and therefore not 'my fault' as the facilitator), freed me to respond to them as part of the learning process, rather than as if I had personally caused them and needed to apologise and heal them.

Knowing that other people had reflected on and developed approaches to **working with participants' feelings** as part of the group system gave me the confidence to face this under-developed side of my own work, and to research it further—making it a part of my own professional development plan. I continue to ponder the implications of thinking about which **emotions** are the 'most helpful', and what the facilitator's role is and should be in influencing this.

Guidance on how to make the **transfer of learning** more effective led me to plan in many of the steps and techniques outlined below, not only during workshops but also in the pre- and post-workshop phases.

Techniques I use in workshops to bridge the gap between knowing and doing

Emotional responses are bound to be generated by the subject matter of sustainability, and are important in motivating people to change in the face of dominant trends. It seems to me that there are two basic ways of taking account of this in our practice:

- Reflecting on and discussing the irrational, emotional responses explicitly with the group

- Trying to induce particular emotional responses that will be most effective in catalysing change

I am uncomfortable with the latter course; it may be manipulative (unethical) and clumsy (unsuccessful). I have concentrated on building my capacity to facilitate discussion about barriers to change, including those created by people's emotional responses.

It has taken some courage for me to begin conversations about emotions with a group; I have been afraid of not being able to cope with others' emotions, especially if I am assuming that it is my role as the facilitator to rescue the person from their emotion. Advanced group facilitation training helped me to recognise this fear, and develop strategies for overcoming my reticence.[8] I am aware of strong emotion and of denial as regular and reasonable phenomena. I no longer see it as a personal failing (or success) and I do not argue with other people's emotions.

There are a number of techniques with which I now feel comfortable.

Direct questions

Asking an open question to the group: *'How are you feeling?'*

Kick-starting a round of disclosures, or paired discussion: *'I think this would be a good time to share what you feel about what we have been learning.'*

Responding to a participant's display of emotion: *'Jo(e), you sounded sad/angry/ enthusiastic then, can you tell us about what you're feeling?'*

Spectrum lines

This needs a room with space for everyone to line up along a wall or an imaginary line, one end being the 'not at all' end, and the other being the 'extremely/completely' end. Once the question has been asked, participants place themselves along the line. These human sculptures can be used to catalyse a conversation about how people feel about the group's attitude to something. It is important to ask a clear question that does not imply a 'right' answer. For example: *'How likely do you think it is that the world's governments will take action on greenhouse gas emissions?'*

When everyone has placed themselves, the debrief might include questions such as 'how do you feel about where you are on the spectrum?' and 'how do you feel about where everyone else is on the spectrum?'

Engaging the whole person

I use techniques and approaches that engage more than just the intellect:

- Creativity. I use posters, pictures, video, music and props (product samples, games, the natural environment) in my presentations, and I encourage participants to use their own creativity during exercises and assignments

- Ethics. Exercises such as the Earth Evacuation spark discussions about ethics—who should be on the spaceship, and what criteria should be used to decide?

- Physical. Some exercises use physical objects and movement to explain planetary systems

---

8   Led by Mike Eales of Global Resonance in the UK; www.globalresonance.com.

● Imagination. The use of stories and metaphors can illuminate the deeper truths and implications behind environmental and social problems

These approaches help people access their own non-intellectual sides—including their emotional responses—helping to make them explicit and therefore amenable to choice, and also helping to illuminate any contradictions between intellectual and emotional acceptance.

Incisive questions

Incisive questions are designed to replace 'limiting assumptions' (which are self-imposed barriers) with 'freeing assumptions'.[9] This technique can be used to enable people to acknowledge and directly address the barriers created by emotions, which in turn arise from deeply held assumptions. For example:

Facilitator: *'What are you assuming that may be stopping you from acting?'*

Participant: *'Our customers don't value sustainable development when they buy, so there's no point pushing this in my department.'*

Facilitator: *'That may be true or it may not. What else are you assuming?'*

Participant: *'I'm assuming that it's too hard to find customers who do understand, and that if I fail to find them soon, my boss will be angry with me.'*

Facilitator: *'What else are you assuming?'*

Participant: *'I'm assuming that I won't be able to work with my boss to come up with a plan.'*

Facilitator: *'If you knew that you can work with your boss to come up with a plan, what would you do?'*

Participant: *'Well, I'd come up with some proposals for researching and segment-ing our customer base, and talk them through with him. We'd set a timetable and test out the strategy. Now I have a plan!'*

The higher the degree of trust in a group, the more honest people are likely to be in this kind of conversation.

Sphere of influence, sphere of control

It is easy to slip into feeling powerless (and it may be a comfortable place to be). I help people identify their power and feel comfortable with using it, by drawing up the spheres and asking them to think about (and sometimes to list) what they alone have control over, and what they can influence (see Fig. 2.2). This can be done individually, in pairs or in small groups.

Everyone has some ultimate control, even if it is only over whether they switch the lights off when leaving the room. Most people, when they stop to think, have more control than they realise (although exercising it may take some effort—choosing to cycle rather than drive to work). Everyone also has plenty of influence—every conversation is an opportunity to champion sustainability.

9   From the work of Nancy Kline, and developed in detail in Kline 1999. See also www.timetothink. com.

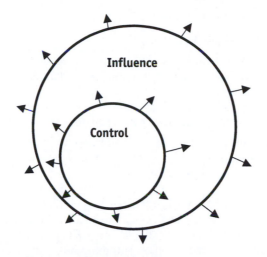

FIGURE 2.2 **Sphere of influence, sphere of control**

Over time, with practice and support, people can extend their spheres of influence and control. Getting an accurate picture of these spheres helps to show opportunities that were hidden, and also helps people see where they can influence but not control, helping them to avoid being made vulnerable to disappointment by their enthusiasm.

It is important to allow people to identify their spheres of influence and control for themselves, rather than the facilitator telling them what they ought to be doing. First, they are bound to know their jobs better than the facilitator does. Second, handing over responsibility to the participants demonstrates the facilitator's trust in their capacity to work it out for themselves; this will carry over beyond the course, when the participant will need to identify opportunities and make decisions without the facilitator. (The facilitator can ask good questions to push them further, but should not make direct suggestions.) Third, this engages and rewards the participants' enthusiasm and motivation.

Conversations around these spheres also include the question of courage: will I be courageous enough to champion sustainability in my workplace?

Reflecting on motivation

The facilitator can ask the group to think about motivation. Focusing on a positive experience (asking 'what keeps you motivated?', rather than 'what demotivates you?') will help people believe that they can motivate themselves and others. Typical responses include:

- Positive feedback
- Nice things happening
- Sunny days

- Getting a response to my enthusiasm

- Doing something about it

A short discussion session such as this can be a useful prelude to a discussion where the facilitator invites the group to commit to action, particularly to supporting each other.

Equipping participants to support each other

One of the key implications of denial for action is, according to Marshall, that 'People will never spontaneously take action themselves unless they receive social support and the validation of others' (Marshall 2001).

A workshop setting, with a group of peers sharing a learning experience, is a good situation in which to generate social support and validation.

Putting time aside near the end of a workshop to look at how to prevent faltering can be invaluable. The initiative can be handed to the participants: small groups can come up with suggestions for keeping the momentum up after the workshop. Suggestions typically include:

- Compiling action lists with names and deadlines, and circulating these after the workshop

- Pairing people up to act as 'buddies', to check progress on action points and provide a sounding board

- Setting times for progress meetings, where participants can celebrate their achievements, identify and solve problems, and set new commitments (it may be appropriate for these meeting to be facilitated)

- Setting up an email group so that participants can stay in touch with each other

- Sending email reminders to people about the commitments they made

- Spending time in the workshop identifying specific barriers, and discussing how to avoid or overcome them

- Setting a time to visit an interesting or enjoyable place as a group—reinforcing a team feeling, and providing informal opportunities to discuss progress

- Making contact with other people in the organisation (or outside it) who will help them change

Researching the client

In some cases, the facilitator will be working with participants from a single client organisation. The clearer a picture the facilitator has of the organisation, the more accurately he or she can guide participants when they are discussing what action they want to take and how they want to change their behaviour. The organisational information that I have found most useful is:

- Who is the sponsor of the learning programme, and what is their level of authority?

- What structures exist in the organisation that might support (or frustrate) participants, and how can these be influenced (e.g. management systems, site committees, performance appraisal schemes)?

- How willing to change is the senior management?

- What are the biggest sustainability issues for the organisation, as judged by itself, and by its stakeholders? How willing is the organisation to address its biggest issues?

- How clear is the organisation on what it wants these participants to do with their new understanding?

If it is not appropriate for the facilitator to research this beforehand, these are questions that participants might want to research for themselves.

How do I judge my success?

The nature of my work makes it hard to evaluate in a rigorous way, although I build in opportunities for clients and participants to give me feedback and I also make time to reflect on workshops and what I have learned from running them. These are some of the comments that have reinforced my own perception of being a more skilled and effective facilitator, since beginning this learning journey.

Overcoming their own and others' reluctance to address sustainable development is highlighted in some people's comments: 'It's a really big help having you here: we talked about things that really needed saying' (project manager, NGO). 'We need a better understanding of what we're talking about—we need to get people talking about it—it doesn't come up as a topic . . . This needs to find its way onto our agenda' (senior managers, chemical company).

The importance of commitment and motivation has been highlighted by participants. 'It's not really about [the company]—I'm committed to making sustainable development happen inside and outside [the company]' (construction engineer).

Participants are sometimes asked to identify the significant, striking or important things that happened during a workshop. Here are two sets of extracts:

Committed communication
Everyone's interested
Level of commitment
More aware and committed
Revitalised initiative
A new dawn!! (We hope)
Re-energised

Enthusiasm from group
Group commitment
Buy-in to project by team

Many ideas extracted from group
The level of enthusiasm of team

More lasting effects are also reported: 'I was very pleased the way the day went, even more when I found that some of the participants were still enthusing about it two days later in a Design Team meeting' (project manager, construction company); and: 'The sales training we planned went ahead, all the European sales team has been trained and people are out there talking to customers about it [sustainability] and winning orders as a result' (sustainability manager, manufacturing company).

What do participants identify as having helped them and their peers to become enthused and achieve these changes? 'Imagining the future success we may create.' 'Thinking forward to what a sustainable future might be like.' 'Collective recognition of the team that we can make a difference.' 'Oh—10 year vision—spooky!' (multi-organisation construction team). 'Interaction, very frank discussion.' 'The basic ideas and the openness and honest way the ideas were presented and discussed' (construction company).

This organisation now has around 60 sustainability champions from all levels of the organisation, who are taking forward a variety of action plans appropriate to their own job responsibilities. There are regular forums for these champions to meet, supporting each other's efforts and problem-solving. Having identified certain internal core management systems as barriers to greater progress, some champions have raised this with senior management and gained support in changing them.

In this case, the courage to be honest, coupled with a critical mass of peers committed to sustainable development, has empowered these junior and middle-ranking managers to confront underlying causes related to core business practices, as well as the more obvious 'housekeeping' issues (paper recycling, bicycle storage) which, while important, only scratch the surface of organisational change for sustainable development.

Conclusions

My experience of teaching sustainability is in situations where the purpose is not purely educational, attending only to someone's level of knowledge and understanding. My work is change management, or even change catalysis, and I have developed my understanding of personal and organisational change to help make this more effective.

To catalyse change for sustainable development, ideally, one would have in place:

● Knowledge and understanding

● Structures and incentives

● Leadership

● Motivation

● Social support and validation

The facilitator can enable knowledge and understanding, motivation, and social support and validation. Sometimes, these three together can be enough to overcome a lack of leadership from above, and a lack of strong external incentives or obligations.

There are some special features of sustainable development which make attending to the emotional responses of participants particularly important:

- The evidence of the crisis can overwhelm some participants with grief, anger or fear

- For others, the evidence of the need for change is at once too terrifying and not immediate enough to their own experience, leading to denial phenomena

- The behaviour change implied by the evidence is at once onerous for the individual, and yet will not make a significant difference to the problem if only that individual acts

- Participants may be returning to workplaces where they will not be supported in transferring their learning

Emotional responses will occur, whether the facilitator intends this or not. The facilitator can choose to incorporate techniques that enable participants to recognise, reflect on and discuss their emotional responses, noticing whether their hearts and their heads are 'on different planets'. Open discussion can free participants from being trapped by their emotions and give them an opportunity to try out having different emotional responses, some of which may be more empowering. The facilitator can also make time for participants to plan together in some detail the action they will take, in effect creating a peer support group—reducing the chances of action being put off back in the workplace.

References

Cambridge Programme for Industry/Forum for the Future (2001) *Sustainability Learning Networks Background Briefings* (Cambridge, UK: Cambridge Programme for Industry).

Galbraith, M.W. (ed.) (1990) *Adult Learning Methods: A Guide for Effective Instruction* (Malabar, FL: Krieger Publishing Co.).

Heron, R. (1999) *The Complete Facilitator's Handbook* (London: Kogan Page).

Hunter, D., A. Bailey and B. Taylor (1996) *The Facilitation of Groups* (Aldershot, UK: Gower).

Kline, N. (1999) *Time to Think: Listening to Ignite the Human Mind* (London: Ward Lock).

Kübler-Ross, E. (1975) *On Death and Dying: What the Dying Have to Teach Doctors, Nurses, Clergy, and Their Own Families* (New York: Macmillan).

Marshall, G. (2001) 'The Psychology of Denial: Our Failure to Act against Climate Change', *The Ecologist*, 22 September 2001.

Roszak, T. (1992) *The Voice of the Earth: An Exploration of Ecopsychology* (Grand Rapids, MI: Phanes Press, 2nd edn): 38.

Taylor, M. (1997) 'Transfer of Learning: Planning Workplace Education Programs', Partnership in Learning, National Literacy Secretariat, Human Resources Development, Canada, www.nald.ca/FULLTEXT/nls/inpub/transfer/English/cover.htm.

Walker, P., and S. Martin (2000) *The Natural Step: A Framework for Sustainability. Training Manual* (Cheltenham, UK: The Natural Step UK).

3
Sustainability in business education
THE LEISURE AND TOURISM CURRICULUM

Shirley Eber

London Metropolitan University, UK

In 1999, the UK Secretary of State for Education and Employment expressed the 'hope that all involved in the education of today's business students will develop and implement sustainable development education strategies'. This chapter is based on a practical project[1] which goes some way to realising the government's 'hope'.[2] It shows why and how sustainability can and should be integrated into business education, focusing on the undergraduate tourism curriculum. In today's world, sustainability is an important aspect of business in general and tourism in particular, and its integration into the teaching of both these areas is discussed here. Such discussions are central to the project, in order to persuade those 'involved in the education of today's business students' who are yet to be convinced of the need for developing and implementing 'sustainable development education'.

1 The project is a partnership between Tourism Concern, a UK-based non-governmental organisation campaigning for fair and equitable tourism, and the Business School at London Metropolitan University; it is funded by the Department of Environment, Food and Rural Affairs (Defra).
2 In 1998, the UK government set up the Sustainable Development Education Panel (SDEP) in order to identify gaps and opportunities in the provision of sustainable development education. The subsequent HE21 Project, carried out by the Forum for the Future (1999), surveyed the status of sustainable development education in business curricula, and involved 25 partner universities that recognised the need to incorporate more specific sustainability learning in some professional and vocational areas; London Metropolitan University Business School identified the leisure and tourism (L&T) curriculum as one such area.

Tourism makes an ideal case study for such a project, for at least two major reasons: the global nature of tourism, and its increasing popularity as a vocational area of study. Tourism is a truly global business. According to the World Tourism Organisation (WTO), it is the world's largest industry, surpassing those such as automotive products and chemicals.[3] In 1999, international tourist arrivals reached 664 million, generating receipts of US$455 billion. In its report to the WSSD (World Summit on Sustainable Development) in Johannesburg in 2002, where tourism was on the agenda, the WTO forecast that tourism is likely 'to triple over the next two decades, with nearly 1.6 billion visiting foreign countries by the year 2020'.[4] The rapid growth of tourism—with the attendant implications as a sector of potential employment and economic activity—is reflected in the mushrooming of tourism studies. In the UK, the number of institutions offering undergraduate tourism degrees grew by 450% in the 1990s (Airey and Johnson 1998). That tourism is indeed a significant business is confirmed by the fact that the majority of these courses (54%) are taught within 'business, tourism or service sector management departments' (Airey and Johnson 1998).

Tourism is also an activity that illustrates the importance of sustainability, since it clearly epitomises the complex interactions between the economic, environmental and social (including cultural and political) spheres. Evidence from around the world shows that the unplanned and uncontrolled growth of tourism increases 'the pressure on the natural, cultural and socio-economic environments of popular destinations' which in turn requires 'more intensified efforts to address sustainability in tourism development.'[5] Today's tourism students—the future industry leaders—urgently need to be conversant with sustainability debates so that they can enter the business equipped to be at the forefront in moving this important sector towards a more sustainable world.

Sustainability

The concept of sustainable development was embraced and promoted by the Brundtland Report (*Our Common Future*; WCED 1987) in 1987, acknowledged by the international community at the Rio Summit in 1992 (in the Agenda 21 set of guidelines), and taken up again at the WSSD in Johannesburg in 2002. Throughout these years, it has been much debated at UN and international fora, by academics, scientists, NGOs, public-sector development agencies, and private business. However, most emphasis has been on *environmental* sustainability, since the global effects of mis- or overuse of the Earth's natural resources have become increasingly apparent. But, even in its initial formulation, the concept of sustainable development was always broader and more far-reaching, stressing the interrelationship of the environment with economy and society: the 'triple bottom line' (Elkington 1998). This broader agenda was echoed in 2001 by Margaret Beckett of the UK government's Department for Environment, Food and Rural

---

3   www.world-tourism.org, accessed 2001.
4   *Ibid.*
5   *Ibid.*

Affairs (Defra), who stated that 'government support for sustainable development goes beyond concerns for the environment';[6] and, even more succinctly: 'sustainable development is about social justice' (Starkey and Welford 2001: xxviii).

The past decade has witnessed a proliferation of protests against a variety of global and local events, such as global warming, flooding, and food and agricultural crises. Demonstrations against globalisation, the North American Free Trade Agreement (NAFTA) or the World Trade Organisation (WTO) negotiations have brought together activists, academics and students from developed and developing countries, and have seen anarchists marching alongside militant socialists and Christian human rights groups. However disparate their causes may seem, the demonstrators are united by a discontent with the unsustainable way in which the world is ordered. The explicit and implicit demands of the participants are for greater transparency, responsibility and accountability on the part of national and international institutions, governments and businesses, and include calls for social justice and human rights, fair trade and equity, poverty alleviation, and the protection of natural and cultural diversity. After meeting with anti-globalisation protestors in February 2002, Belgian prime minister Guy Ver-hofstadt concluded that: 'Their message is that globalisation needs a political counterpart to tackle the social, ecological and cultural consequences of the world becoming one economy' (Hari 2002)—what better definition of sustainable development could there be?

Sustainability and business

Businesses are beginning to understand that they can no longer ignore the clamour for change. Indeed, sustainability has appeared 'on business conference agendas, then [become] the subject of seminars in [its] own right' (Egan and Wilson 2002: 49). Thus, multinational corporations, such as the oil giant Shell, widely condemned a few years ago for its decision to sink the Brent Spar oil rig, are now claiming that sustainability is integral to their operations:

> Human Rights. None of our business? Or the heart of our business? Profits and principles: is there a choice?
>
> Human Rights. It's not the usual business priority. And for multinational companies operating in developing countries, it could be tempting to dismiss it; to call it a socio-political issue rather than a business one, and hope it just goes away.
>
> At Shell, we are committed to support fundamental human rights and have made this commitment in our published Statement of General Business Principles. It begins with our own people, respecting their rights as employees wherever they work in the world. We invest in the communities around us to create new opportunities and growth. And we've also spoken out on the rights of individuals—even if the situation has been beyond our control. It's

6   BBC World Service, 27 June 2001.

part of our commitment to *sustainable development* [author's emphasis], balancing economic progress with environmental care and social responsibility.
In today's business environment, we don't pretend there are any easy
answers, but we continue to stay involved. Because making a living begins
with respecting life.[7]

A plethora of organisations are engaged in campaigning to encourage socially
responsible, environmentally sound and economically viable business practice.[8] Examples include the launch (in spring 2001) by the highly respectable Financial Times
Share Index (FTSE) of indices for socially responsible investment (SRI) to 'identify companies with the strongest records of corporate social and environmental performance'
using 'globally recognised standards to define best practice in corporate responsibility'.[9] The FTSE initiative was created in association with the Ethical Investment
Research Service which aims to enable investors 'to make informed and responsible
investment decisions'.[10] Some of the other business organisations and consultancies
include: SustainAbility, focusing on 'how the sustainable development agenda fits
within business strategy';[11] AccountAbility, which is 'committed to enhancing the performance of organisations and to developing the competencies of individuals in social
and ethical accountability and sustainable development';[12] and the UK Social Investment Forum, a network for 'socially responsible investment, including ethical, green
and cause-based investment'.[13]

Some cynicism about the use of the term 'sustainability' in business is undoubtedly
healthy. Professor Erik Cohen has described sustainability as a fashionable concept
which 'can easily be adopted by entrepreneurs to advertise their tourism product, without any real steps being taken to apply it' (Cohen 2002). Whether the adoption of sustainability by business is interpreted as a marketing ploy—'greenwashing' for competitive advantage—or seen as a genuine change in management practice is a moot point.
What is clear, however, is that such corporate moves come in response to the dissatis-
faction and demands of consumers, shareholders and stakeholders. Thus, according to
Simon Zadek (2001: 1):

> business is in the limelight of increasingly concerned public scrutiny. The
> popular media carries daily fresh allegations of its misdemeanours. An out-
> pouring of books, pamphlets, films and conferences challenge and debate its
> social and environmental performance . . . recent years have seen the emer-
> gence of the philosophy and practice of 'corporate citizenship'. Corporations
> have sought under this umbrella to gain broader trust and legitimacy through
> visibly enhancing their non-financial performance. Today, the focus is shift-
> ing from philanthropy to the impact of core business activities across the
> broad spectrum of social, environmental and economic dimensions repre-
> sented by the vision of *sustainable development* [author's emphasis].

7 Shell advertisement, *New Statesman*, 7 October 2002, www.shell.com.
8 A preliminary search has revealed over 30 UK-based websites dealing with aspects of sustainability in business.
9 www.ftse4good.com, accessed 2001.
10 www.eiris.org, accessed 2001.
11 www.sustainability.co.uk, accessed 2001
12 www.accountability21.net, accessed 2006
13 www.uksif.org, accessed 2001

The extent to which business and sustainability are compatible is also the subject of lively debate, epitomised in January 2002 when the former director of Greenpeace, Lord Melchett, took up an advisory post with PR company Burson Marsteller. The firm's 'core business', according to environmental activist George Monbiot, is 'defending companies which destroy the environment and threaten human rights' (Monbiot 2002). While Melchett's intention may be to help big business adopt corporate social responsibility, Monbiot views his move as no less than an act of betrayal and part of the threat facing 'environmentalism . . . [which] is in danger of being swallowed up by the corporate leviathan'.

These examples highlight just some of the complex issues with which students need to grapple and between which they need to discriminate if they are to understand the contexts and contents of the sustainability debates. This understanding forms a basis that they can bring to an employment market in which the demand for 'sustainability-competent' graduates is undoubtedly growing.

Sustainability and the tourism business

In the tourism business, perhaps more so than in other business sectors, companies small, medium and large are espousing adherence to sustainability principles. Major corporations such as British Airways and Intercontinental Hotels, prestigious awards such as Tourism for Tomorrow, business associations such as the Association of Independent Tour Operators, the World Travel and Tourism Council, and non-governmental organisations such as Tourism Concern,[14] are just some of the bodies advocating greater responsibility in the tourism business. Internet searches have uncovered over 20 tourism-related websites dealing with sustainability. As with other sectors, the extent to which some of these tourism business initiatives are mere marketing, public relations exercises or 'greenwashing' rather than commitment to genuine sustainability policy and practice is debatable.

Likewise, definitions about what constitutes sustainability in tourism abound among various market segments and academics. Terms such as 'ecotourism', 'nature-based tourism', 'responsible tourism', 'alternative tourism', 'green tourism' and 'ethical tourism' are often used interchangeably with sustainable tourism (Dawson 2001). The World Tourism Organisation's interpretation, however, encapsulates a broad triple-bottom-line approach:

> Sustainable tourism development meets the needs of present tourists and host regions while protecting and enhancing opportunities for the future. It is envisaged as leading to management of all resources in such a way that economic, social and aesthetic needs can be fulfilled while maintaining cultural integrity, essential ecological processes, biological diversity and life support systems.[15]

14   Tourism Concern's publication *Beyond the Green Horizon* (Eber 1992) was the first multi-stakeholder discussion paper to set out principles for sustainable tourism.

15   www.world-tourism.org, accessed 2001.

To turn this statement into practice, the World Tourism Organisation has set a 'frame of reference for the responsible and sustainable development of world tourism' through a global ten-point Code of Ethics for Tourism, drawn up and approved in 1999 after extensive consultation with over 70 of the organisation's member states.[16] The code is the first of its type to include an enforcement mechanism and clearly highlights the linkage between ethical considerations and sustainability.

Teaching sustainability in business

While sustainability is increasingly finding its way into business practice, it largely remains a cinderella topic within undergraduate business teaching in the UK. The subject benchmark statement for bachelors' degrees in General Business and Management issued by the Quality Assurance Agency for Higher Education (2000)[17] only includes sustainability (as well as business ethics and globalisation) among 'contemporary and pervasive issues' listed at the end of other, presumably more important, areas to be addressed. However, some headway is being made at secondary-school level with the introduction of Sustainable Development and Citizenship as subjects within the National Curriculum. Such developments lay the groundwork and provide excellent opportunities for extending the debates into tertiary education.

Why there is a dearth of sustainability in undergraduate business teaching may be explained by a report carried out for Defra (2000: 1), which identifies the problem:

> There is little understanding of the current language used to communicate Sustainable Development and hence limited education about the issues it encompasses. It is not only the general public that is not speaking the language; even those whose job it is to work in the various fields of sustainability struggle to communicate effectively with the language currently available to them.

Delivering sustainability requires not only understanding 'the current language', but moving from traditional subject-based teaching towards a more integrated, holistic approach. If educators themselves—and those designing and delivering teacher training and development programmes—face difficulties in grappling with sustainability issues, it follows that they too will 'struggle to communicate effectively' with their students. This misunderstanding explains a certain resistance on the part of some teaching staff, noted within this project, towards proposals to integrate sustainability, which they regarded as the imposition of a 'politically correct', value-laden agenda, and a limitation on their academic freedom. Many such reservations were dispelled by exploring the issues in close consultation with staff at all levels (see 'The way forward' below), during which it became clear that an essential element of a sustainability approach is the examination of competing values, encouraging critical analysis and debate.

Some teachers also doubted the capacity of undergraduate students, particularly in their first or second years of study, to grasp complex debates. However, our own teach-

16   www.unwto.org/code_ethics/eng/global.htm, accessed 14 February 2007.
17   www.qaa.ac.uk, accessed 2001.

ing experience has shown that students are keen to engage with current and important issues when helped to see how relevant they are to their present lives as consumers and citizens, and to their potential futures as employees, employers, managers and entrepreneurs. Indeed, many young people, 'who have grown up with an acute awareness of environmental issues' (McConnell 2002), seem to be well ahead of their teachers. A recent survey carried out by the National Union of Students (NUS) found that 'three-quarters of student job hunters would not work for a company with a poor ethical record and half of those surveyed would take less pay to work for a company with a good history' (McConnell 2002).

Teaching sustainability can be made more accessible to students by eliciting relevant issues and themes from their own experiences, and developing them in a progressive fashion. The challenge lies in finding ways of exploring sustainability issues at an appropriate level without detracting from their complexities. If this is already being done in secondary-school education, it is certainly possible to accomplish at tertiary level. The solution lies in student-centred learning, using young people's own concerns as a starting point from which to explore the debates.

Teaching sustainability and the tourism business

Tourism in higher education in the UK is overwhelmingly taught within business and management schools. In 2001, of 145 academic institutions offering degrees in tourism (or leisure and tourism), 106 were in business or management[18] reflecting the fact that tourism in the real world is primarily a business. However, it is disappointing to note that only 19% of the prospectuses for these courses 'specifically refer to the inclusion of the "fashionable" topic of "sustainable" tourism', the relative absence of sustainability being 'closely linked to the extent to which the course is vocationally driven' (Airey and Johnson 1998: 14).[19] The apparent mismatch between sustainability and vocational aims is due, according to John Tribe (Botterill and Tribe 2000), to the influences that bring about a 'potential contest' over the design of tourism curriculum. These influences are those that promote it

> as a vocational one for commercial ends [including] the needs of employers, professional bodies, academics rooted in business departments . . . [and] influences which promote the tourism curriculum as one for non-commercial ends. For example host and environmental interests would promote a curriculum for sustainable and responsible tourism, and academics from critical subjects will promote a more open agenda for tourism studies (Botterill and Tribe 2000: 5).

18   Universities Central Admissions Service (UCAS 2001); www.ucas.com, accessed 2001.
19   Internationally, attempts to address this absence include a think-tank, organised by the Business Enterprises for Sustainable Travel (BEST) in February/March 2001, aimed 'to produce a set of model curricula on the subject of Sustainable Tourism to be distributed . . . to educational institutions . . . throughout the world'; www.sustainabletravel.org, accessed 2001.

The dichotomy between vocational and non-commercial aims in tourism education is confirmed by the Quality Assurance Agency for Higher Education (QAA) which accepts that most tourism degrees 'still lay emphasis on career and vocational objectives' (Botterill and Tribe 2000: 12). However, the agency does recognise that most tourism 'programmes have broadened from their vocational origins to embrace wider issues relating to the nature, impacts and meanings of tourism, thereby furnishing an understanding of what is now a major world phenomenon' (Botterill and Tribe 2000: 12). Although the QAA's subject-specific guidelines for tourism education do not use the term 'sustainability', its outline for desired learning outcomes do include some relevant elements, in that a tourism graduate will be able to demonstrate an understanding of

> the role of tourism in the communities and environments that it affects and in particular:
>
> - . . . the relationship between tourism and the communities and environments in which it takes place
>
> - . . . the contribution and impacts of tourism in social, economic, environmental, political, cultural and other terms
>
> - . . . the approaches to managing the development of tourism through concepts of policy and planning
>
> - . . . the ethical issues associated with the operation and development of tourism (Botterill and Tribe 2000: 14).

Since tourism is an activity, industry or business that is highly dependent on the quality of natural, social and cultural environments, it is hardly surprising that the majority of degree courses include sessions covering its impacts. However, dealing with the impacts of tourism does not equate to the comprehensive and holistic approach that characterises sustainability. Mowforth and Munt (1998) describe the clear distinction: rather than dealing with 'the environmental, economic and socio-cultural impacts *of* tourism', a sustainability approach entails 'seeking to understand how socio-cultural, economic and political processes operate *on* and *through* tourism' (Mowforth and Munt 1998: 2). Thus, where an impacts analysis lists 'the outputs or consequences of tourism', a sustainability approach provides a context in which 'tourism is seen as a focal lens through which broader considerations can be taken into account' (Mowforth and Munt 1998: 3).

Parameters of sustainability in tourism teaching

Tourism itself is a highly diverse activity, organised by a multi-sectoral industry, entailing a range of economic and cultural transactions within a variety of social and political contexts. As a field of study, this very complexity calls for a multidisciplinary approach, drawing on subjects such as 'economics, environmental theory, social theory, politics, geography and international relations' (Mowforth and Munt 1998: 3)—all of which come within the remit of a sustainability approach. Thus the teaching of tourism as a business must focus on 'sustainability and tourism business' within the broader generic context of 'sustainable business', rather than on the more limited topic of 'sustainable tourism'. In other words, students need to grasp the broad principles, contexts and debates of sustainability in business, and then learn how these apply to the business of tourism (and, indeed, to related subjects such as travel, leisure, hospitality, events management, the arts and cultural industries).

The three model units or modules outlined below have been designed to provide such a programme. These 12-week modules are not meant to be isolated, stand-alone units but taught within a curriculum revised to embed sustainability (see 'The way forward' below). The contents are graded in a progressive fashion, though how they are organised and implemented will vary according to the overall design, structure and focus of programmes at various academic institutions.[20] However, we believe that the units provide a framework for and can be adapted to particular pathways or specialisations in *any* business curriculum.

At first level the module entitled 'Business in Society' introduces and explores basic issues concerning sustainability in business in general. Students should understand and appreciate:

1. **Problems of being unsustainable:** economic, environmental, social, cultural and political dilemmas facing our planet which need to be addressed by business, governments and individuals

2. **Global and local business:** nature and dimensions of business and interrelationships with trade, politics and development issues

3. **Introduction to sustainable development:** backgrounds to, major interpretations and definitions of the concept of sustainable development

4. **Sustainable development and the public sector:** agendas of public-sector national and international bodies and non-governmental organisations, and their roles in sustainable development

5. **Sustainable development and the private sector:** agendas of private-sector national and international businesses, shareholders and stakeholders, and their roles in sustainable development

6. **Corporate social responsibility (CSR):** the business case for sustainability and corporate approach to responsibility, community participation and consultation

7. **Citizenship, responsibility and ethics:** the rights, responsibilities and ethical considerations of individuals—as citizens, consumers or stakeholders—towards global and local sustainability issues

8. **Social justice and human rights:** human needs, human rights, poverty alleviation and social inclusion in sustainable development

9. **Environmental dimensions:** actions and responsibilities of businesses, NGOs and individuals in protecting global and local environments

---

20 At London Metropolitan University, each session is delivered as a lecture which is followed up and reinforced through student-led group seminars.

10. **Diversity, culture and heritage:** globalisation of business and relationships to the diversity of cultures, notions of heritage, identity and community

11. **Moving forward:** options for the future—constructive engagement or opposition, genuine change or competitive marketing, ethical careers

12. **Plenary:** overview of main issues raised

At second level, the module entitled 'Tourism Business in Society' develops and extends the issues covered at first level in the context of the business of tourism (or other business pathway), with emphasis on relevant and illustrative case studies. Students should analyse and apply:

1. **Problems of unsustainable tourism:** the unplanned and uncontrolled growth of international tourism and its economic, environmental, social, cultural and political consequences

2. **Global and local tourism business:** interrelationships between global and local trade in tourism, politics and development

3. **Introduction to sustainable development of tourism business:** backgrounds to and major definitions and interpretations of concepts of sustainable tourism

4. **Sustainable development of tourism and the public sector:** roles, agendas and actions of public-sector national and international bodies and NGOs in the sustainable development of tourism

5. **Sustainable development of tourism and the private sector:** roles, agendas and actions of private-sector national and international tourism businesses, shareholders and stakeholders in the sustainable development of tourism

6. **Corporate social responsibility (CSR) in tourism:** the business case for sustainability in tourism, and the approaches and actions of tourism companies towards their responsibilities to the environment and communities

7. **Citizenship, responsibility and ethics in tourism:** the rights, responsibilities and ethical considerations of individuals—as tourists and stakeholders—towards global and local tourism issues

8. **Social justice and human rights in tourism:** the role of tourism in issues of human needs, human rights, poverty alleviation and social inclusion

9. **Environmental dimensions of tourism:** the actions and responsibilities of tourism businesses, NGOs and individuals in protecting global and local environments

10. **Diversity, culture and heritage in tourism:** relationships between the globalisation of tourism business and the diversity of cultures, interpretations of heritage and notions of identity

11. **Moving forward:** options for the future—constructive engagement or opposition, genuine change or competitive marketing, and ethical careers in tourism

12. **Plenary:** overview of main issues raised

At final level, the module entitled 'Solutions, Tools and Techniques for Sustainability' builds on topics explored at first and second levels, and provides the bases of skills applicable to the real world of business, emphasising case studies and practical field work. Students should apply, critically appraise and evaluate:

1. **Life-cycle and carrying capacity analysis:** patterns of tourism development; physical, ecological, and social carrying capacity models and techniques (e.g. limits of acceptable change [LACs], land-use planning and zoning)

2. **Sustainability indicators:** methods to assess natural and cultural diversity, resource use, waste, pollution, local production, human needs, access to facilities and decision-making in relation to sustainable development of destinations

3. **Area protection:** status and control of protected areas, e.g. national parks, wildlife reserves, sites of special scientific interest (SSSIs), etc.

4. **Environmental auditing, impact analysis (EIAs) and reporting:** techniques (e.g. rapid rural appraisal [RRA], geographic information systems [GISs]); role of environmental auditing, impact analysis and reporting in tourism

5. **Social auditing, impact analysis and reporting:** techniques for evaluating, planning for and mitigating the influences of tourism on the social environment of destinations

6. **Cultural auditing, impact analysis and reporting:** techniques for evaluating, planning for and mitigating the influences of tourism on the cultural environment of destinations

7. **Product stewardship and responsibility:** nature and development of the tourism product; needs, goals and responsibilities of industry, destination communities and tourists in relation to sustainable development of the tourism product

8. **Stakeholder analysis, engagement and conflict resolution:** agendas of different stakeholders in tourism (governments, 'host' communities, industry representatives, NGOs and tourists) and the role of mediation in consensus building

9. **Consultation and participation techniques:** range of consultation/participation techniques; levels of participation, ownership and control, representation and community institution building

10. **Sustainability legislation, policy and control mechanisms:** role of legislation versus self-regulation for industry, (local, national and international) governments and professional associations (e.g. Agenda 21) in changing tourism practice

11. **Codes of conduct:** range and status of codes for tourism industry, tourists, host governments and communities

12. **Monitoring and assessment overview:** environmental, social and cultural performance indicators to monitor and assess sustainability in tourism development

The way forward

This project has gone some way towards realising the integration of sustainability within undergraduate tourism degrees taught at London Metropolitan University, and has produced a guidelines paper (Eber 2002) for dissemination to other academic institutions. The following section summarises the methodology used in the current project for those interested in adopting similar objectives.

Two broad strategies emerged as the most successful way of embedding sustainability into the curriculum: first, a review of existing course contents in order to integrate sustainability into existing core modules and ground it within and across the curriculum; and, second, the introduction of discrete core sustainability modules (generic and indicative) in order to ensure the maximum exploration of the issues and debates. Implementing these complementary strategies provide a curriculum with a coherent and progressive direction.

The review of course contents was carried out by first mapping sustainability elements within existing modules and subsequently identifying gaps and opportunities for its introduction. This was done by means of a detailed questionnaire distributed to and completed by teaching staff. The questionnaire aimed to highlight sustainability concepts and solutions already being taught, to identify sustainability concepts and solutions which staff consider relevant to their modules, and to assess the feasibility of integrating sustainability concepts and solutions into their modules. It proved important to interview respondents after they had completed the questionnaires in order to clarify responses and debates arising from them. These discussions helped to allay any misunderstandings and reservations staff had concerning the sustainability 'agenda'.

To support both staff and students in teaching and learning new sustainability contents, appropriate bibliographies and other resources were identified. A staff seminar and workshop were held for teachers in recognition that additional training is necessary to deliver sustainability. Students are encouraged to assess sustainability aspects in their work placement experiences (between second and final years), in their final dissertations, and helped to identify suitable post-graduation employment.[21]

Finally, sustainability is included within course aims and objectives, learning outcomes, assessment criteria, and prospectuses and publicity. This ensures that everyone—academic staff at all levels and students—is clear about and agrees with the overall approach.

---

21 'People and Planet' advises graduates on ethical careers: www.ethicalcareers.org, accessed 2001.

It is clear that 'many people, not least young people, are seeking greater involvement in the decisions which affect their lives' (SDEP 2001). Students need to be knowledgeable about and equipped to deal with these decisions as they step out into an increasingly flexible and insecure world of business and work. The transferable skills that are increasingly advocated in higher education—including the ability to think critically, communicate effectively, handle and interpret information, appreciate and act in the context of social and cultural diversity, and make ethical evaluations[22]—are all enhanced and developed through a sustainability approach. The overall aim is to produce globally and socially aware graduates with the necessary skills and knowledge to respond to the 'emergent new business agendas of the 21st Century'[23] which have been discussed above. There can be little doubt that 'there are both scope and need for even more engagement, for students in higher and further education, if all are to be encouraged towards a greater commitment to democratic, sustainable communities' (SDEP 2001). It must surely be the role of educators to provide students with knowledge as broad and comprehensive as possible which equips them for business as informed citizens of an increasingly global world. This chapter has demonstrated how students can be encouraged in this direction within the field of tourism business education; we believe that other fields can and must follow suit.

References

Airey, D., and S. Johnson (1998) *The Profile of Tourism Studies Degree Courses in the UK: 1997/98* (Guideline no. 7; National Liaison Group for Higher Education in Tourism [NLG]).

Botterill, D., and J. Tribe (2000) *Benchmarking and the Higher Education Curriculum* (Guideline no. 9; National Liaison Group for Higher Education in Tourism [NLG]).

Cohen, E. (2002) 'Authenticity, Equity and Sustainability in Tourism', *Journal of Sustainable Tourism* 10.4: 267-76.

Dawson, C. (2001) 'Ecotourism and Nature-Based Tourism: One End of the Tourism Opportunity Spectrum?', in S.F. McCool and R.N. Moisey (eds.), *Tourism, Recreation and Sustainability: Linking Culture and the Environment* (Wallingford, UK: CABI Publishing): 41-53.

Defra (UK Department for Environment, Food and Rural Affairs) (2000) 'Towards a Language of Sustainable Development', www.defra.gov.uk/environment/sustainable/educpanel/language/01.htm, accessed 2001.

Eber, S. (1992) *Beyond the Green Horizon: Principles for Sustainable Tourism* (Godalming, UK: Tourism Concern/WWF).

—— (2003) *Integrating Sustainability into the Undergraduate Curriculum: Leisure and Tourism* (Guidelines No. 10; Association for Tourism in Higher Education; www.athe.org.uk).

Egan, J., and D. Wilson (2002) *Private Business, Public Battleground: The Case for 21st Century Stakeholder Companies* (London: Palgrave).

Elkington, J. (1998) *Cannibals with Forks: The Triple Bottom Line of 21st Century Business* (Oxford, UK: Capstone Publishing).

22　These skills are based on the core capabilities embedded across the undergraduate curriculum at London Metropolitan University.

23　University of North London (now London Metropolitan University) Business School Strategic Plan 1996/97–2000/01.

Forum for the Future (1999) *HE 21 Sustainable Development Education: Business Specification* (London: Department for Environment, Transport and Regions [DETR]; www.forumforthefuture.org.uk).

Hari, J. (2002) 'An Unlikely Supporter', *New Statesman*, 25 November 2002.

McConnell, S. (2002) 'Pouring oil on troubled waters', *The Guardian*, 20 July 2002.

Monbiot, G. (2002) 'The business of betrayal', *The Guardian*, 15 January 2002.

Mowforth, M., and I. Munt (1998) *Tourism and Sustainability: New Tourism in the Third World* (London: Routledge).

SDEP (Sustainable Development Education Panel) (2001) *Annual Report* (London: SDEP, March 2001).

Starkey, R., and R. Welford (eds.) (2001) *The Earthscan Reader in Business and Sustainable Development* (London: Earthscan).

UCAS (2001) 'Guide to Tourism-Related Courses, 2001', www.ucas.co.uk, accessed 2001.

WCED (World Commission on Environment and Development) (1987) *Our Common Future* ('The Brundtland Report'; Oxford, UK: Oxford University Press).

Zadek, S. (2001) *The Civil Corporation: The New Economy of Corporate Citizenship* (London: Earthscan).

4

Getting it

UNDERSTANDING THE SCIENCE AND PRINCIPLES BEHIND SUSTAINABILITY

Darcy Hitchcock and Marsha Willard

AXIS Performance Advisors Inc., USA

Learners new to the concept of sustainability need a grounding in its definition and the underlying scientific principles. Teaching the science, even to non-scientists, is not that difficult. Overcoming misconceptions, prejudice, disinformation and guilt are the real challenges. This section will address these specific obstacles to learning.

Challenges

Unmuddling the term 'sustainability'

'Sustainability' as a label has some significant drawbacks. First, people often misconstrue it to mean sustaining their own organisation. It's a complex concept, fuzzy over specifics, and to some seems too intellectual and abstract to have much value. Also, 'sustaining' something rarely has the same emotional appeal as creating something; sustainability has a feel of levelling out the carnage instead of inspiring a 'renewal economy' or some other term. But, for better or worse, in most situations it's the best label we have to describe the concept.

In some situations, you're better off using another term altogether, a bridging concept that gets you most of what you want. For example, in manufacturing, 'zero waste' might resonate well. Companies that have had a strong total quality programme may find it easiest to frame this as 'enlarging the definition of quality to include the environment (and perhaps also society)'. See Box 4.1 for other options.

In describing sustainability, we often use the 'three Es': Economy, Environment, and social Equity. We talk about how these are often traded off in our society instead of designing a way for them to be collectively optimised. We use a systems diagram to make this point. The heat island effect is the tendency for urban areas to be significantly warmer than the surrounding countryside due to their heat-absorbing surfaces. This leads to a downward cycle in all three 'Es'.

We explain that sustainability is about turning these gears in the opposite direction, coming up

- Zero waste
- Green building
- Green chemistry
- Community health
- Social responsibility
- Triple bottom line
- Resource efficiency
- Product certification
- Enlarged definition of quality

Box 4.1 Alternatives to 'sustainability'

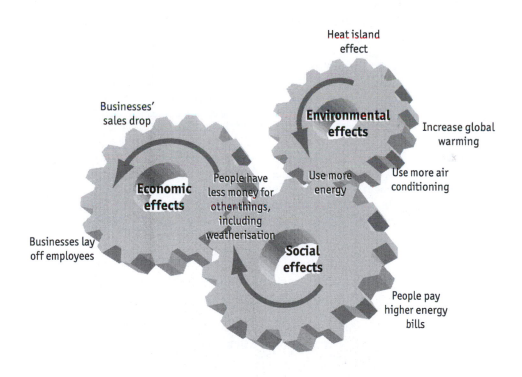

FIGURE 4.1 **The downward cycle of the three 'Es'**

with solutions that simultaneously improve the economy, the environment, and the health of our society (see Fig. 4.1).

Organisations often confuse sustainability with other 'green' or environmental programmes. In service organisations, when we try to explain the importance of sustainability, we often get a blank stare and 'But we recycle our paper . . .'. So we use the following diagram to help people distinguish sustainability from the other, necessary but lower forms of corporate programmes (see Fig. 4.2).

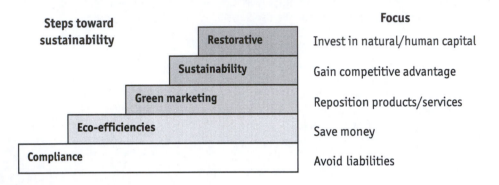

FIGURE 4.2 **Steps toward sustainability**

On the low end are organisations that are still focused only on regulatory compliance. Organisations at this level are primarily concerned with avoiding legal liabilities and may view environmental issues as a source of additional costs and headaches. Organisations focusing on eco-efficiencies have discovered that saving resources not only helps the environment but also their financial bottom line. So they are interested in saving water, energy, raw materials, etc. Here the focus is internal. At some point, many organisations realise that being 'green' can attract new customers or make their community more attractive. At this point they use green marketing to differentiate themselves from others.

Both green marketing and eco-efficiencies focus on 'doing better'. However, when organisations understand sustainability, they begin to wonder, 'Are we doing enough?' Just 'doing better' may not be enough to live within the limits of nature. When organisations reach the level of sustainability, they understand in their hearts the need to significantly change what they're doing and also understand in their heads the incredible business opportunity this can afford. Some businesses go beyond even sustainability (which balances our demands on nature with what it can provide) to restoration, rebuilding what we have degraded.

Last, we find it is often useful to translate sustainability into the logical responsibilities the organisation should assume. We use the following worksheet for businesses to explain their responsibilities.

Moving toward sustainability—worksheet

Document your thoughts about how to change your organisations practices to become more sustainable.

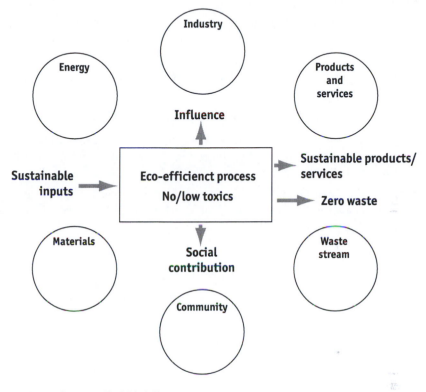

FIGURE 4.3 Input/output 'bubble' diagram

We talk through an example from another industry. For example, what might a restaurant do over a ten-year period? It could :

- Buy green power and organic produce (sustainable inputs)
- Use waste heat from its ovens to heat water to wash table cloths and other linens (eco-efficiencies)
- Find alternatives to toxic cleaners (low toxics)
- Donate all leftovers to homeless shelters, compost all their food scraps (zero waste and social contribution)
- Advocate for fair wages and living conditions for migrant labour (industry influence)

Using an example like this makes sustainability seem attainable.

Dealing with negative perceptions associated with environmentalism

While many people may view themselves as environmentalists, many others have negative associations with the environmental movement. So it can be important to distance sustainability from these negative connotations. Here are tactics we have used to avoid this problem:

● Pitch sustainability to their own interests. You can emphasise the economic benefits to business people and human interests with social service organisations

● Avoid using common environmentalist terminology: ecosystems, earth, planet, etc.

● Avoid blitzing them with doomsday statistics about the environment

● Openly acknowledge the mistakes that some in the environmental movement have made: being adversarial, appearing to care more about other species than people, etc.

● Explain how sustainability is different from past 'green' efforts: it's non-political, collaborative, doesn't place blame, acknowledges the need for a healthy economy and profitability, etc.

Getting past 'science phobia' and technical jargon

Not everyone enjoyed their science classes in high school, and certain audiences can be turned off by technical, scientific jargon. In such audiences, it's important to find terms, examples and demonstrations that connect with their experience. As an example, the following instructions explain how we present The Natural Step (TNS) System Conditions to a lay audience, including concepts of the evolution of life on earth, bioaccumulation, entropy, etc. (see Box 4.2). (A summary of the four System Conditions can be found on page 97.)

Explaining The Natural Step System Conditions

Materials

A picture of the Earth that you can write on; we use the view with Europe, Africa and Asia; props for the evolution example (ammonia, soda bottle, whoopee cushion or picture of natural gas flame, metals, oxygen mask, dinosaur, etc.), props for entropy demonstration (coffee mug and sugar cube, or fishbowl and food colouring), overheads of all System Conditions and the listing of chemicals in breast milk.

Process

Put up the picture of the Earth. Explain that there are three ways we humans have changed the Earth.

Box 4.2 Explaining The Natural Step System Conditions (continued opposite)

Explain that one way is that we take things out of the Earth's crust. Draw an up-arrow or shovel near the North Sea. Explain that the atmosphere was not always like it was, and to put things into historical perspective you need a volunteer to hold out their left arm.

Do the evolution demonstration where the shoulder is the beginning of life on Earth and the tip of the fingernail is today. What was the atmosphere like for the first half? (Heavy metals precipitated out; first life-forms could 'eat' ammonia, carbon dioxide and methane and released oxygen so, by the elbow, the ozone layer was forming. More carbon was sequestered even up to the middle of the palm when dinosaurs reigned. Humans showed up somewhere along the fingernail and the dust on the tip of the nail is the entire Industrial Revolution.) Ask: 'So what did we do in the Industrial Revolution to change this picture?' Answer: 'We dug all this stuff back up again!'

Show and discuss SC#1.

Go back to diagram of the Earth and explain that another way we have changed this picture is to make things. Draw a factory over Eastern Europe.

Show the overhead of chemicals. Ask: 'What do you think this list represents?' Explain that it's in the breast milk of women and polar bears. Ask: 'How did it get there?'

Explain bioaccumulation: Imagine a farmer sprays a pesticide onto a field so that every leaf of grass has the tiniest drop on it. As the cow eats the grass it ingests this chemical and, because the chemical is persistent (stays around a long time) and the cow's body doesn't know what to do about it and can't use it or excrete it, it stuffs it into fat cells. Over the lifetime of the cow, it eats a lot of the chemical and so has a much higher concentration than the grass. Now you come along and eat a bunch of burgers, and, like the cow, your body doesn't know what to do about this stuff so it stores it in your cells. Over 30 years, you eat a lot of cows. Now you have a higher concentration than the cow. Now you decide to have a baby and you nurse the baby. Who is at the top of the food chain, getting the highest concentration of pesticide? The baby. And this is particularly troublesome because the baby is so small and is growing rapidly.

Show SC#2.

Go back to the diagram of the Earth and explain that the third way we change this picture is to delete portions of nature. Draw a delete sign or cross over Malaysia or Indonesia. Ask: 'How do we delete nature?' Pollution, over-harvesting, development, etc.

Ask why it matters that we delete nature. Discuss the services nature provides. If desired, talk about the lessons from Biosphere 2.

Show SC#3.

Go back to the picture of the Earth and explain that, if we didn't do these things, we would have a sustainable society and these principles are derived from scientific principles around which there is no disagreement. But the scientists recognised a need for a social condition. Draw a sad face over Africa. Show SC#4 and discuss fairness and efficiency. (If in your presentation you did not go over the funnel, this is a good time to do the population demonstration. You can also talk about the Ecological Footprint.)

Box 4.2 (from previous page)

Motivating reluctant learners

It's possible that some participants in your presentations or classes may not be there willingly. So it's important to find ways to engage them. We have found the following strategies helpful:

- When people introduce themselves, ask them to mention a problem in their community that worries them. These will almost always fit into one or more of the three Es of sustainability: Economy, Environment, and social Equity (or livability). Make the connection for them so they can see sustainability as a way of solving a problem that they care about

- Plan lots of activities and group interaction to keep it interesting

- Introduce humour where possible

- Use lots of stories and examples. These draw people in

Managing guilt and defensive behaviours

One of the biggest psychological barriers is despair and helplessness. Many people feel guilty about their lifestyles and choices; they know about the environmental problems. But they don't know how to fix it so they push it all away. They may also be overwhelmed with the plethora of environmental/social issues—global warming, ozone hole, pesticides, cancer rates, species extinction, drought, etc., etc. Missing the overall relationship between these problems, it may feel like too many plates to keep spinning in the air, especially when kids need to go to soccer practice, the cupboards are bare, and they have an upcoming business trip out of town. Here are some ways to handle some of these challenges:

- Declare a guilt-free zone at the beginning of your presentation. Explain that none of us is sustainable. Explain that sustainability shouldn't be about not having what you want; it should be about getting what you want sustainably

- At the beginning of your presentation, ask people to list environmental problems that worry them. Write these on a flipchart. Make a long, messy list. Add your own if they stall out or need prompting. Now say that all these problems can be tied back to a handful of mistakes we made in designing our society. I then go on to explain The Natural Step (TNS) System Conditions. You can go back to their list afterward and show how all theses disparate problems can be tied back to these four principles

- Include an activity that shows them how to apply the System Conditions to everyday decisions. (See the activity in Box 4.3. I often include a worksheet that forces them to compare each option to each System Condition.)

Avoiding arguments about the data

Closely associated with guilt and defensive reactions, we have, on occasion, had people who refuted the basic data (e.g. the pace of species/topsoil/rainforest decline,

Everyday actions

Make up a set of cards, cutting out pictures from magazines to represent each of the options. Then ask the students to discuss each option in light of each System Condition and select the best and worst option based on The Natural Step framework.

- You are at the grocery store to buy juice
  - Buy frozen juice (–)
  - Buy fruit and make your own juice (+)
  - Buy fruit juice (not from concentrate)
- You are landscaping part of your yard
  - Plant a western red cedar tree (+)
  - Plant grass (–)
  - Plant a tulip tree
- You are choosing a sport for exercise
  - Play golf (–)
  - Play baseball
  - Do snow shoeing (+)
- Your family is deciding on a vacation
  - Go on a canoe trip in the Cascades (+)
  - Fly to Hawaii
  - Take a cruise (–)
- You are buying cleaning products
  - Buy Orange Plus (+)
  - Buy Lysol in an aerosol can (–)
  - Buy 409 in a spray bottle
  - Buy Simple Green
- You are thirsty
  - Drink bottled water
  - Grab a soft drink (–)
  - Get water from tap (+)
- You are choosing personal care products
  - Buy Tom's of Maine
  - Buy Dry Idea (–)
  - Buy deodorant stone (+)
- You are picking what to eat at a restaurant
  - Order beef: a steak or hamburger (–)
  - Order chicken
  - Order spaghetti with marinara sauce (+)

Box 4.3 Everyday actions (continued over)

● You are choosing a hobby
  – Get into photography
  – Hike/backpack (+)
  – Go fishing in motorised boats (–)
● You're at the checkout counter at the grocery store
  – Ask for paper bags
  – Ask for plastic bags (–)
  – Bring your own canvas bag (+)
● You want breakfast
  – Eat oatmeal (+)
  – Eat cold cereal
  – Eat a frozen waffle (–)

Box 4.3 (from previous page)

whether global climate change is actually happening, etc.). In these situations, it's important not to get into a my-data-is-better-than-your-data argument or to embarrass the individuals raising the issues. We have found the following strategies can help you get out of these quagmires:

● Agree with as much as you can with what the person said (e.g. acknowledge that there is still a lot of uncertainty about the rate of species decline, etc.)

● Explain the sources for your data. Ideally, you are using widely accepted sources for your data (e.g. the UN Intergovernmental Panel on Climate Change). In these situations, I will mention that the media, in their attempt to tell 'both sides of a story', often don't do a good job of explaining that one perspective is backed up by, for example, 2,500 of the world's leading climatologists, and the other represents a handful of scientists paid by industry

● If there is any reason to suspect the data you are referencing are not entirely objective, acknowledge that. For example, non-profits can benefit by making things look worse than they are just as industry groups may have an interest in making things look better. In these situations, you can often go on to say, 'Let's assume for a minute that their estimates are overstated by a factor of two. That still doesn't eliminate the basic problem. We're still not sustainable; the rate of degradation is just slower.'

● Express interest in reviewing their sources. Tell them once you've had a chance to look it over, you'd be glad to comment. In some cases, I pass these by experts in the field and forward their comments

What follows in the first two chapters of Part II are two activities that flow from the above discussion. The first activity is the Sustainability Card Game, which aims to anchor the science behind the four System Conditions of TNS. The second activity out-

lines the challenges that learners face in taking the message back to their organisations. It helps them learn how to effectively sell sustainability in their organisations by
addressing these challenges.

Part II
Hands-on exercises/ activities

5
The Sustainability Card Game

Darcy Hitchcock and Marsha Willard

AXIS Performance Advisors Inc., USA

Purpose

This activity anchors the science and TNS System Conditions, helping people to make important distinctions.

Preparation

You will need one set of cards per table (4–6 people). Copy the cards onto coloured paper, preferably card stock, and then cut them into cards. (*Tip:* If you print each set onto different-coloured paper, it will make it easier to sort them if they get mixed up.) Shuffle each deck of cards so that each table doesn't start with the same card (which can create a distracting 'echo effect' in the room.) You can use the science or the System Condition cards separately. Feel free to remove cards that you think aren't as relevant to your needs or create new ones. If you don't have much time for this activity, 'stack the decks' so the most important cards are on top. If desired, you can have the players keep score by retaining cards they answered correctly.

To conduct

Put a deck of cards on each table. Tell the participants to have one person pick up the first card and read the question. The person opposite them (or anyone at the table) can try to answer the question. After a short discussion, the person holding the card should read the answer that is printed below the question. Then, the next person at the table should pick up the next card. Tell them how much time you have allocated for this activity and encourage them not to take too much time on any one card; instead, try to get through most of them. If there are any cards they don't understand or want to talk about more, have them set them aside to discuss during debriefing. After giving the instructions, rove around the tables to make sure they understand the instructions and to answer questions.

To debrief

Ask each table if they had any cards they set aside to discuss in more detail (e.g. they didn't understand the answer, they didn't agree with the answer, they wanted more information about it, etc.) If none of the tables has any cards set aside, ask them to tell you which cards generated some of the biggest surprises or best insights. If the group is still reticent, you can mention some of the issues you saw them struggling with when you roved around the tables. If you had the participants keep score, hand out a prize to those who got the largest number of questions correct.

Time

To go through all the cards, allow 30 minutes plus an additional 15 for debriefing. If time is short, you can split one deck of cards, giving each table just a portion of the deck.

Science question #1

When we throw something away, what are we really doing and which scientific principle is involved?

A: There is no 'away' in a closed system; we're just moving things around. (Matter and energy cannot be destroyed.)

© 2002 AXIS Performance Advisors

Science question #2

When you drive your car around, eventually the gasoline in the tank disappears. What happens to it and which scientific principle is involved?

A: In the process of burning the gasoline, the chemical bonds are broken, creating energy. The elements in the gasoline are mostly turned into gases attached to oxygen (e.g. carbon dioxide, carbon monoxide, sulphur dioxide, etc.) (The first and second laws of thermodynamics: matter and energy cannot be destroyed; and entropy—things disperse.)

© 2002 AXIS Performance Advisors

Science question #3

People often say that trees make oxygen. Is this true or false and why? Which scientific principle supports your answer?

A: False. Trees don't make oxygen; they release it from its bond with carbon atoms, from carbon dioxide. The trees keep the carbon to grow new limbs and leaves, giving off the oxygen as a waste product. (Matter and energy cannot be destroyed.)

© 2002 AXIS Performance Advisors

Science question #4

Entropy implies that, in a closed system, matter and energy spread and devolve over time. Why haven't the Earth's systems stopped functioning?

A: The sun powers the Earth's systems, causing the ocean and wind currents and fuelling plant growth (green cells). Without the sun, the Earth would quickly become a frozen, dead wasteland. Without green cells, most life on earth would die. Photosynthesis is the primary way in which nature organises dispersed matter.

© 2002 AXIS Performance Advisors

Science question #5

Entropy implies that everything spreads over time. How do metals spread, leaking into nature?

A. They rust (e.g. iron), evaporate (e.g. mercury), or may be leached (e.g. lead batteries in landfills).

© 2002 AXIS Performance Advisors

Science question #6

Entropy implies that everything spreads over time. How do petrochemicals spread, leaking into nature?

A: We burn them (e.g. gasoline) or they off-gas (e.g. vinyl). Oil tankers may have accidents (e.g. *Exxon Valdez*), and they pump bilges, which allow oil residue to spread by ocean currents.

© 2002 AXIS Performance Advisors

Science question #7

How could chemicals such as DDT get into polar bears, even though the chemicals may never have been used in the arctic region?

A: These chemicals can be distributed by ocean and wind currents. They may flow into streams, into the ocean and then travel up the food chain. They may migrate with birds, butterflies or whales. Weather patterns and the tendency of organic chemicals to evaporate except in cold climates tend to move these chemicals to the poles. They bioaccumulate and biomagnify in the fat of fish and animals there.

© 2002 AXIS Performance Advisors

Science question #8

Since matter cannot be destroyed, when you eat a carrot, what do you actually consume?

A: You consume the structure and concentration of matter (the stored energy) in the carrot. You absorb some of the elements to fuel your own body. Digestion and other biochemical processes break down the chemical bonds, releasing energy. The rest of it is excreted.

© 2002 AXIS Performance Advisors

Science question #9

When we mine gold and refine it into ingots, we increase the economic value. Is this an example of improving the (1) order, (2) structure or (3) concentration of matter?

A: Concentration. The ingot is a more pure form of the gold than what is mined.

© 2002 AXIS Performance Advisors

Science question #10

When a jeweller forms gold into a lovely necklace, the economic value of the gold has been increased. Is this an example of improving the (1) order, (2) structure or (3) concentration of matter?

A: Structure. The artistic process has changed how the gold looks and how it can be used.

© 2002 AXIS Performance Advisors

Science question #11

When we combine iron and other metals to make steel, we have increased the economic value. Is this an example of improving the order or concentration of matter?

A: Order. The elements are rearranged into new molecules, making them stronger than iron itself.

© 2002 AXIS Performance Advisors

System Condition question #1

What are three materials that we get from the Earth's crust in a manner that violates System Condition #1?

A: Fossil fuels, metals and minerals.

© 2002 AXIS Performance Advisors

System Condition question #2

What are two ways that human activities violate System Condition #3?

A: Over-harvesting (e.g. fishing, timber), displacement (e.g. development, dams) or manipulation (e.g. agricultural practices such as genetically modified organisms).

© 2002 AXIS Performance Advisors

System Condition question #3

Does the first System Condition imply that we can't use metals?

A: If we could recycle all or almost all of metals so that they didn't leak into nature faster than the natural rate at which they are redeposited, we would avoid violating the first System Condition. We can also use abundant metals (such as aluminium) as long as they don't accumulate in nature.

© 2002 AXIS Performance Advisors

System Condition question #4

Oil can be used for many things. What is probably the most damaging thing we do with oil and why is it a problem?

A: We burn it, releasing carbon dioxide and other greenhouse gases, contributing to climate change.

© 2002 AXIS Performance Advisors

System Condition question #5

Why were the eagles more affected by the pesticide DDT than the fish that they ate? What is the scientific term for this process?

A: Bioaccumulation and biomagnification. DDT accumulates in tissue (bioaccumulation). The concentration of DDT increased as it moved up the food chain, so the eagles had much more of it in their body (adjusted for body weight) than each fish they ate (biomagnification).

© 2002 AXIS Performance Advisors

System Condition question #6

If your electricity comes mostly from coal-burning power plants, which System Condition are you predominately violating?

A: SC#1, since coal is mined from the Earth's crust at a much faster rate than it is redeposited in the crust.

© 2002 AXIS Performance Advisors

System Condition question #7

CFCs, which are destroying the ozone layer protecting us from the sun's radiation, are a violation of which System Condition?

A: SC#2, since these chemicals are human-made (synthetic) and cannot be processed by natural systems.

© 2002 AXIS Performance Advisors

System Condition question #8

System Condition #1 involves the use of metals. If you had an application where you could use any of the following three metals, which would be the best to use from an environmental standpoint and why: gold, copper or aluminium?

A: Aluminium is a prevalent metal in nature and human-caused flows of it are dwarfed by nature's. The other metals are more rare. In general, the more rare and heavy the metal, the more potentially harmful it is to nature.

© 2002 AXIS Performance Advisors

System Condition question #9

When a farmer sprays a slurry of manure on fields, such that the excess flows into rivers, what happens and which System Condition is most violated?

A: The excess nitrogen from the manure-based fertiliser contributes to eutrophication of water bodies, where the excess nutrients leads to algae blooms which deplete the dissolved oxygen. Then fish and other organisms die off. Even though manure is biodegradable, making too much of it is a violation of System Condition #2. Factory farms can be considered human-made. You may also consider this a violation of System Condition #3 because it degrades natural systems.

System Condition question #10

Some people support the use of nuclear power because it doesn't generate greenhouse gases. Which System Condition(s) could you use to argue against the use of nuclear power?

A: Radioactivity comes from materials in the Earth's crust, violating System Condition #1. And they decay so slowly, taking tens of thousands of years. Plutonium is not found in nature; it is refined from uranium, so it also relates to System Condition #2. You can also use System Condition #4. We should be promoting technologies that will help people around the world meet their basic needs. Do we really want every country on Earth to have a nuclear fuel source that requires special handling and could be used for weapons?

System Condition question #11

What are three ways in which our current agricultural practices violate System Condition #3?

A: (1) Monocultures reduce diversity; (2) fields displace habitat; (3) soil is routinely depleted.

System Condition question #12

When you choose a vegetarian meal over one with beef, pork or chicken, which System Conditions are you impacting?

A: Animal products often involve a lot of antibiotics and hormones (System Condition #2); they require a lot of water and land, both for themselves and their feed (System Condition #3); and we could feed more people using the grain directly, versus feeding it to the animals (System Condition #4).

System Condition question #13

When you choose to let your lawn go brown in the summer instead of watering it, which System Conditions are you contributing to?

A: By using less water, you reduce the need for new dams and reservoirs (System Condition #3). Since there is a limited and finite amount of fresh water available to meet human and other needs, you are also contributing to System Condition #4.

6

The Multi-organisation Business Simulation

Darcy Hitchcock and Marsha Willard

AXIS Performance Advisors Inc., USA

Selling the message: enlisting support for implementing sustainability in your organisation

Once people are excited about sustainability, they face the tough job of selling it in their organisations. Experience teaches that passion for the subject is not only insufficient in converting an organisation, it is often counterproductive if people view the missionary as over-zealous. This chapter outlines a simple exercise that prepares learners to take the message back to their organisations and to help them learn how to effectively sell it in their organisations by addressing these challenges.

Challenges

As outlined earlier, the challenges that the learners will face are:

- Establishing the urgency and relevance of sustainability
- Soft-selling with hard data
- Overcoming resistance
- Getting top-management support

- Building critical mass

- Preparing a strategy

- Making the abstract actionable

- Managing the complicated trade-offs in implementing sustainability

- Identifying appropriate starting places

- Aligning existing business systems to support sustainability

- Overcoming the 'immune response' in an organisation

- Dealing with external barriers (vendors, community, etc.)

The following exercise will give learners an opportunity to find ways of dealing with these challenges.

The Multi-organisation Business Simulation

Purpose

To give people an opportunity to apply sustainability and The Natural Step System Conditions in a complex, realistic, but hypothetical situation.

Preparation

In this business simulation, there are four businesses which operate in the same town: a silicon wafer plant, food processor, medical centre and a company manufacturing period lighting products. (There are other businesses represented on the site map as well.) Set the room up with four tables. Each table will represent a different organisation. If you have more than four tables, you'll end up with more than one table playing each of the companies. If you have fewer than four tables, choose the most relevant organisations to represent. For each table, provide enough copies of the instructions for each person and several copies of the organisation's two-page handout (people can share).

To conduct the activity

Hand out the materials and explain that they are to review the materials given to them about the organisation. Walk through the instructions worksheet. They are to come up with ways to make their organisation more sustainable. Press them to be creative. Just being a little greener won't be enough. Tell them how much time they have to complete the exercise and then rove around the tables to make sure they are on track. (If time allows, you may choose to do this exercise in stages, where the tables complete each step, report, complete the next step, report out, etc. You can also modify the instructions sheet to meet other goals as well. For example, we've used this simulation to make distinctions between a pollution prevention approach and true sustainability.)

To debrief

Ask someone from each table to report to the larger group: what company they represented; their best quick-win/big-win ideas; any 'ah-ha's' (insights) they learned from the activity. Ask the other tables if they can think of anything else. Add any comments you may have about what real companies have done.[1]

Time

This is an elastic activity that can take several hours or only 30 minutes. In our experience, an hour is usually ideal: 40 minutes for them to work at their tables and 20 minutes to debrief.

Sustainability Business Simulation: handouts

Instructions

Read the background on the organisation you have been assigned. Then, as a group, use backcasting to come up with recommendations to become more sustainable:

1. Identify your environmental impacts (as best you can), using the four System Conditions to organise your data. Pick a couple of high-priority areas (see Fig. 6.1)

2. Decide what you would consider a sustainable level for each of the high-priority areas (see Fig. 6.2)

3. Brainstorm changes you could make to this organisation to make it more sustainable. From the brainstormed list, identify at least one 'quick win' (something that could be done easily without a lot of time or money) and a 'big win' (something that would make a huge difference in your sustainability performance but which would take significant time and/or money) (see Fig. 6.3)

---

1   There are opportunities to sell waste-streams to other businesses in the community (see Fig. 6.4). The washing water from the food processor could irrigate the farmer's fields. The quartz from the silicon wafer plant can be used by the gravel company. The silicon slurry may be used by the concrete company. Excess hot water from the silicon wafer factory might pre-heat autoclaves for the medical facility.

System Condition 1: From the Earth's crust	**System Condition 2:** Human-made substances
System Condition 3: Habitat	**System Condition 2:** Human needs

FIGURE 6.1

Priority impact	Sustainable level

FIGURE 6.2

Quick win	Big win

FIGURE 6.3

Good Snacks, Inc.

Background

You and the others in your team represent the leadership from a snack food company (primarily corn and potato chips [crisps]) which sits on the outskirts of town. You employ 900 people and expect to grow by 10% over the next two years. (Site map and manufacturing process diagram are attached; see Figs. 6.4 and 6.5.)

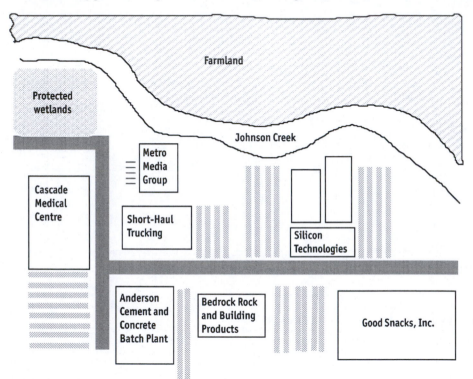

FIGURE 6.4 Site map (this is handed out to each table)

You have recently been presented with the principles of sustainability and are convinced that this is an issue that won't be going away any time soon. You have convened the leadership team to discuss the issue and the implications to your business. While you have never been out of compliance with environmental regulations, you are beginning to believe that this is not enough.

Process diagram: Good Snacks, Inc.

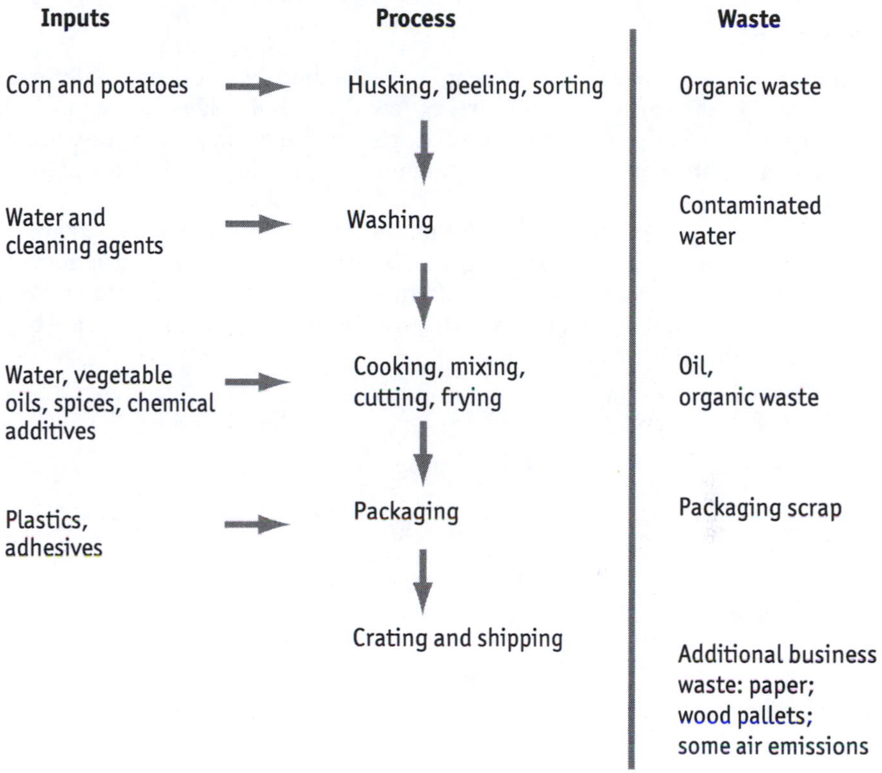

Inputs	Process	Waste
Inputs	**Process**	**Waste**
Corn and potatoes →	Husking, peeling, sorting	Organic waste
Water and cleaning agents →	Washing	Contaminated water
Water, vegetable oils, spices, chemical additives →	Cooking, mixing, cutting, frying	Oil, organic waste
Plastics, adhesives →	Packaging	Packaging scrap
	Crating and shipping	Additional business waste: paper; wood pallets; some air emissions

FIGURE 6.5

Silicon Technologies

Background

You and the others in your team represent the leadership from a silicon wafer manufacturing facility that sits on the outskirts of town. You sell wafers to computer chip makers around the world. You employ 900 people and expect to grow by 10% over the next two years. (Site map and manufacturing process diagram are attached; see Figs. 6.4 and 6.6.)

You have recently been presented with the principles of sustainability and are convinced that this is an issue that won't be going away any time soon. You have convened the leadership team to discuss the issue and the implications to your business. While you have never been out of compliance with environmental regulations, you are beginning to believe that this is not enough.

Process diagram: Silicon Technologies

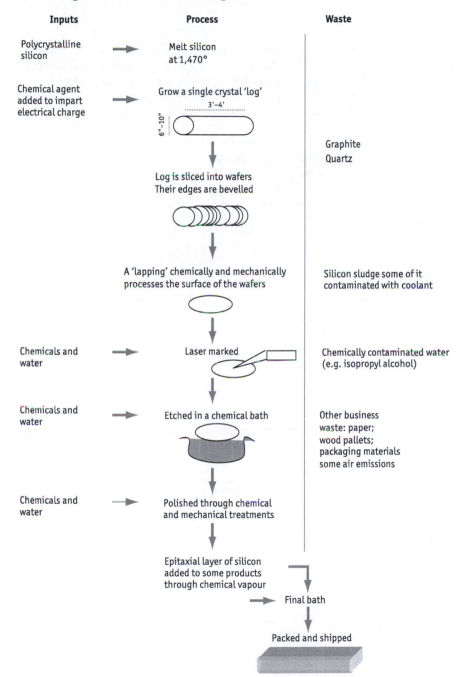

Inputs	Process	Waste
Polycrystalline silicon	Melt silicon at 1,470°	
Chemical agent added to impart electrical charge	Grow a single crystal 'log'	Graphite Quartz
	Log is sliced into wafers Their edges are bevelled	
	A 'lapping' chemically and mechanically processes the surface of the wafers	Silicon sludge some of it contaminated with coolant
Chemicals and water	Laser marked	Chemically contaminated water (e.g. isopropyl alcohol)
Chemicals and water	Etched in a chemical bath	Other business waste: paper; wood pallets; packaging materials some air emissions
Chemicals and water	Polished through chemical and mechanical treatments	
	Epitaxial layer of silicon added to some products through chemical vapour	
	Final bath	
	Packed and shipped	

FIGURE 6.6

Metro Media Group

Background

You and the others in your team represent the leadership from a public relations and marketing business which sits on the outskirts of town. You develop promotional campaigns and produce supporting media (print, video, web page design) for businesses and non-profit organisations. You employ 65 people and expect to grow by 10% over the next two years. (Site map and business process diagram are attached; see Figs. 6.4 and 6.7.)

You have recently been presented with the principles of sustainability and are convinced that this is an issue that won't be going away any time soon. You have convened the leadership team to discuss the issue and the implications to your business. While yours is not a regulated industry and tends not to be considered a major 'polluter', you are beginning to realise that this does not mean that sustainability is an irrelevant business issue.

Process diagram: Metro Media Group

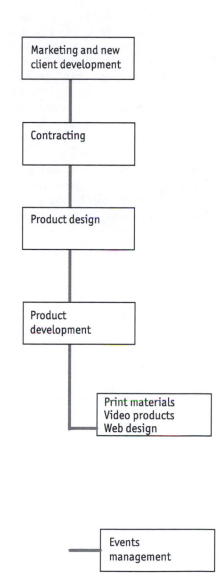

Process

Waste

Marketing and new client development

General office waste: paper; copier and printer toner; office supplies

Contracting

Product design

Product development

Print materials
Video products
Web design

Paper
Inks
Film processing waste

Events management

Food waste
Food accessories (disposable utensils, cups, plates, napkins, etc.)
Serving platters
Print materials (invitations, programmes, etc.)
Event give-aways (door prizes, favours, etc.)
Decorations

FIGURE 6.7

Cascade Medical Centre

Background

You and the others in your team represent the leadership of a small regional hospital. You serve most of the medical needs of the semi-rural community in which you are located. Your facility sits on the outskirts of town in an industrial park. You employ about 300 people. Your facility provides the only cancer and trauma treatment in a 50-mile radius. Your operations also include a medical lab which serves local physicians as well as the hospital and an ambulance service. Because of your location, many of your clientele are Medicare and Medicaid recipients. People looking for affordable housing are causing a precipitous rise in the population of your area. You expect to serve an increasing number of people over the next five years. (Site map and business process diagram are attached; see Figs. 6.4 and 6.8.)

You have recently been presented with the principles of sustainability and are convinced that this is an issue that won't be going away any time soon. You have convened the leadership team to discuss the issue and the implications to your business. While you have never been out of compliance with environmental regulations, you are beginning to believe that this is not enough.

Process diagram: Cascade Medical Centre

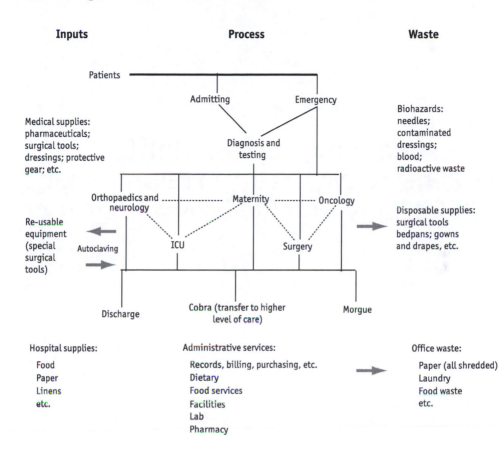

FIGURE 6.8

7
Personalising sustainability
AN INTERACTIVE ACTIVITY TO REINFORCE THE PRESENTATION OF THE NATURAL STEP (TNS)

Joshua Skov

Good Company, USA

Introducing the concepts of sustainability to any audience means presenting both intellectual and emotional challenges. For almost everyone, there are information gaps to fill in, and big-picture concepts to provide. But information is not enough: it important for everyone to feel a personal link between his or her behaviour and the health of society and the planet. The challenge in presenting the concept of sustainability lies in the complexity of our interdependence with each other and the planet.

This chapter presents an activity for helping an audience to understand that relationship between sustainability issues and individual behaviour. Using a straightforward, single-page handout for each person, the exercise presented herein leads a group through the mental process of tying the System Conditions of The Natural Step (TNS) to several commonplace individual behaviours.

Sustainability must be understood personally and emotionally, as well as generally and conceptually. The Natural Step framework can be a foundation for deeper personal and intellectual understanding of important issues, and this activity uses TNS System Conditions as a springboard for engaging the framework.

The activity succeeds for three distinct reasons. First, it gently and effectively removes the attention from the presenter/lecturer and places it on the audience members, individually and collectively. Second, in less than an hour, it provides learning in both individual and group modes. And third, it requires individuals to internalise the links between TNS System Conditions and everyday activity, by presenting 'findings' (i.e. the information from and insights based on their respective handouts) to each other in a non-threatening, small-group setting. This learning-by-teaching leads the participant to internalise that knowledge and take the other participants' individual teachings more seriously.

The Natural Step: a tool for connecting to the ideas of sustainability

For those readers not already familiar with TNS, this section briefly describes the origins and content of TNS System Conditions. For readers who have already attended a TNS workshop or presentation, it may still be interesting to hear of the framework's history.

The Natural Step is a simple conceptual framework that describes high-level conditions by which life on the planet must abide in order to survive in the long run. It was created as the result of a collaborative effort spurred by Karl-Henrik Robèrt, a Swedish oncologist who became troubled by the increasing frequency of child cancers. Robèrt, frustrated by the medical profession's exclusive focus on minutiae in the face of daunting large-scale trends, assembled a group of scientists to come up with a simple framework for understanding the basic tenets of biological survival.

The group used basic science as its starting point, focusing on the Earth as a (nearly closed) system, and using fundamental concepts such as the laws of thermodynamics. After many iterations, the group generated the following four 'System Conditions': that is, requirements for long-term sustainability on Earth, from the perspective of human activity.

In order for a society to be sustainable:

- **System Condition 1.** Nature is not subject to systematically increasing concentrations of substances extracted from the Earth's crust

- **System Condition 2.** Nature is not subject to systematically increasing concentrations of substances produced by society

- **System Condition 3.** Nature is not subject to systematically increasing degradation, destruction and over-harvesting

- **System Condition 4.** Resources are used fairly and efficiently to meet human needs globally

These conditions attempt to capture all physical impacts on and interactions with the natural world. The conditions are not helpfully *prescriptive*, but they provide a framework for understanding impacts as diverse as mineral extraction, declining biodiversity, climate change, deterioration of the ozone layer and the loss of coral reefs.

This framework is straightforward, and even intuitively obvious to many observers at first blush. However, because of its simplicity, it is flexible: it allows people with varying tolerances of detail and with different levels of knowledge to engage the material at whatever level they choose.

Activity summary, materials and set-up

The essence of the activity is simple: hear TNS System Conditions, apply the knowledge and share insights. After hearing The Natural Step framework, workshop participants must examine a particular everyday activity (buying a pair of shoes, driving to work, drinking a latté, etc.) in light of the four System Conditions.

Except for handouts (one per person) and writing utensils, there is no additional set-up required beyond the venue itself. However, the activity is a little easier if participants have a place to write easily, such as tables or desks.

Example handouts appear as an appendix to this chapter. However, I encourage people to make new handouts, to refine these and others, and to share what they come up with, since there are countless examples. Also, since we constantly get new information about the negative impacts of our lifestyles, and since more benign or even restorative alternatives constantly appear, this new information should appear in the handouts whenever possible.

Implementation and sequence

The format can of course be adjusted, and in fact the basic idea is quite flexible. I suggest the following gradual 'roll-out' to maximise the effect:

1. Present an introduction to sustainability, including TNS System Conditions. Hand out the stapled activity page. Instruct participants not to open it

2. Ask the participants to take 3–5 minutes of individual time to make notes linking the action to the TNS System Conditions, using only the one-sentence description as a guide (without opening the paper)

3. Open up the paper (i.e. unstaple it and reveal the facts about the action) and read the contents

4. Allow 5–10 minutes of individual time to make notes linking each of the various facts with one or more TNS System Conditions

5. Allow 10–15 minutes of small-group time to present individual examples to the group and discuss the issues

I have found this format to be effective for a variety of reasons. First, it puts some responsibility on the individual to learn about a particular issue and—most important—to then communicate the knowledge to others.

Second, after the presentation of the framework (which often takes a lecture format), the gradual revealing of the facts keeps individuals engaged because sustainability sneaks up on them and turns out to be relevant to their lives. Often, the one-sentence action is not obviously linked to what the participants think 'sustainability' is.

The format also allows individuals at different knowledge levels to participate throughout. For many people, the TNS principles make intuitive sense at first sight, and examples of each may be immediately obvious. More often, the System Conditions seem big, broad and vague. The close examination of a single, apparently mundane activity provides both insight and a sense of relevance. This feeling of relevance expands in the group activity, as other everyday activities are linked to broader issues of sustainability.

Here is the format I have used most often, in slightly greater detail:

1. Distribute the stapled-shut activity sheets *before* you go over the TNS System Conditions; ask participants to read the one-sentence action, leave the paper stapled shut and then set the paper aside. Then present an introduction to sustainability as you would otherwise do for a workshop, including TNS System Conditions with some detail, examples and question-and-answer time. (This works well because participants usually forget about the action on their sheets)

2. Ask the participants to take 3–5 minutes of individual time (no interaction among participants) to make notes linking the action to the TNS System Conditions, using only the one-sentence description as a guide (i.e. not opening the paper)

3. Ask the participants to open up the paper (i.e. unstaple it and reveal the facts on the action). Ask the participants to read through all of the facts once before they continue

4. Have the participants take 5–10 minutes of individual time (no interaction among participants) to make notes linking each of the various facts with one or more TNS System Conditions. Tell the participants that a given fact may relate to more than one System Condition, and that they should note when this is the case

5. Allow 10–15 minutes of small-group time (3–5 people per group) to present individual examples to the group and discuss big-picture issues. Important note: Because there are 5–8 different action-handouts, each group should have no more than one of each activity

Example sequence

Let us consider one action statement and trace its use throughout the sequence of the activity. (The complete worksheets for seven actions appear in the appendix.) It is helpful to see it in terms of the sequence described above.

We will consider participant Jane Doe. She finds the action 'I drank a latté' on her activity sheet.

1. Just before the presentation of The Natural Step, stapled-shut activity sheets are passed out. Jane's says, 'I drank a latté'. 'Interesting,' Jane says to herself, 'I actually *did* drink a latté this morning . . .' As instructed, she leaves the

paper stapled shut and sets it aside. She then sits back and enjoys a 45-minute presentation of The Natural Step. Within a few minutes, she has forgotten about the activity sheet.

2. At the end of the discussion of the four System Conditions, the workshop facilitator asks the participants to take a few minutes of individual time to consider the activity sheet again, and to make notes linking the action to the TNS System Conditions. 'Oh yeah,' Jane thinks, 'the bit about the latté. Now how does that fit in to all of this?'

Using the information on the four System Conditions that she has just learned, Jane jots down the following items on her paper (without unstapling it):

SC 1: can't think of anything . . .

SC 2: can't think of anything . . .

SC 3: cutting down trees to make paper cups—not necessarily sustainable

SC 4: low wages for poor coffee farmers, not equitable

3. The facilitator then asks participants to open up the paper; Jane does so, and sees a list of facts about her action (i.e. about coffee; see Box 7.1)

Coffee production (overwhelmingly for export) supports 20–25 million farmers and agricultural workers worldwide. Coffee exported to the United States comes primarily from South and Central America. (Coffee is the second most traded good after oil, measured by the total value of international trade.)

Coffee production traditionally has consisted of 'mixed-shade' systems in which coffee is cultivated together with trees grown for their fruit or wood (for timber or fuel). These mixed-shade systems support high biodiversity—more than any other agricultural system, and nearly as much as some natural tropical forest.

In the past 20 years, production has moved increasingly towards 'sun coffee', intensive monocultural systems that use high-yielding varieties and large amounts of fertiliser and pesticides. These more industrial systems typically have higher short-run yields, but they support very little biodiversity (and often interfere with the routes of migratory birds) and they result in much greater soil erosion and water pollution.

Fertilisers are usually produced from petroleum.

Farmers in developing countries often lack the equipment and training to use fertilisers and pesticides safely. Still, these chemicals are widely available and typically loosely regulated or simply unregulated.

Box 7.1 Facts about coffee

4. At the facilitator's request, Jane takes 5–10 minutes of individual time (no interaction among participants) to make notes linking each of the various facts with one or more TNS System Conditions. 'I see,' Jane thinks, after reading through the information. 'I think I have a few other comments now that I know all that's involved in making coffee!' She manages to come up with something for each System Condition, and she comes up with more than one

item for System Condition 3. She comes up with the following ideas, to add to the ones above:

SC 1: coffee production uses fertilisers, made from oil from the Earth's crust

SC 2: pesticides used in coffee production can be persistent chemicals

SC 3: conventional production destroys bird habitat; may damage water quality

SC 4: cost of coffee (to the drinker) does not include environmental costs!!

5. The facilitator then breaks the large group (about 35 people) into smaller groups to present individual examples to the group and discuss issues. Jane finds that the other four group members have completely different actions, ranging from driving a car to eating a hamburger.

   As each group member presents his or her respective action and its impacts, the others join in with additional ideas. Jane's action promotes much discussion, and the group comes up with the following additional impacts for coffee:

SC 1: most coffee is consumed far from where it's produced so must be transported—usually in a way that's powered by fossil fuels

SC 3: the clearing of wilderness for coffee plantations lowers terrestrial biomass, so it probably contributes to climate change

SC 4: Starbucks makes millions, but poor farmers starve (variation on Jane's original observation)!

   Another member of the group makes the observation that improved practices can have *positive* impacts in terms of the four System Conditions. Three examples might be:

SC 2: organically grown coffee eliminates the use of synthetic pesticides

SC 3: shade-grown coffee provides habitat for migratory birds

SC 4: fair-trade coffee gives a larger share of the retail price to the producer, usually a small farmer in a developing country

As the small-group activities draw to a close, Jane realises that she has applied the System Conditions to a seemingly mundane daily activity, thereby getting a big-picture glimpse of the activity's potential impacts. In the process, she has taught something to her group and learned from their unique perspectives as well.

Additional challenges and tips

Although the activity works well and quite easily once you get it going, there are a few areas in which to be careful, both in creating the activity sheets and in running the activity.

First, it is important to have a balance of information in the activity sheets. For any given action, it is easy to list System Condition violations ad nauseam. However, some people may respond quite negatively to the negative news. As an important aside, research has shown that schoolchildren learn best when bad news is accompanied by a much larger dose of good news simultaneously. Even if you think people need to hear all of the urgent bad news, beware of the possible impact. Remember that if they disengage emotionally, the moment for real learning will be lost and the exercise will have no point.

There are two specific ways to accompany each action with the 'good' aspects of the apparently bad news. Mostly obviously, you can describe the *function* of an activity for human health and well-being. For example, beef is a major source of protein; cars provide transportation for several hundred million people every day. Also, you can use case studies of more sustainable alternatives, such as the comments on shade-grown, organic, fair-trade coffee in the example above. Some people find inspiration in one and not the other, so it is important to provide both.

Second, when people are working in small groups, move from group to group and facilitate participation by as many people as possible. Since everyone has a different handout, everyone has a built-in opportunity to contribute something; take advantage of this asymmetric knowledge to make each person an 'expert' in his or her group.

Third, as the exercise wraps up (usually back in the full group), allow space for discussion of the activity as a whole. Ask them what the activity shows them, facilitate the discussion, and work to bring out some group conclusions or insights. Typically, you will end up with one or more of the following:

- **Each action has some impact in terms of each of the four System Conditions**. Most groups will end up looking deeply at one or more of the handout actions, with the result that the action—whether it is buying shoes or eating a burger—has ripple effects in terms of all four System Conditions

- **To move towards sustainability, we need systemic change**. Most people in every group I have worked with have seen that the big things matter—our houses, our cars, our diet, our commuting patterns, the basic materials of our manufactured goods. Simply put, we cannot simply tweak our way of life. Instead, we need to change the infrastructure and systems of our economy and society

- **There is a non-technical element to sustainability that, for different people, can be philosophical, political or spiritual**. Ultimately, issues of human welfare, fairness and our place in nature emerge in some form in group discussions following this activity. People will express these concerns differently, but the basic idea—that our everyday actions are part of ongoing decision-making about how the world is and should be—is a deep insight that a facilitator or presenter can choose to reinforce. We are part of a system, and sustainability, at its heart, is systems thinking. This activity often gives people a chance to think about the 'system'—a rare opportunity in most people's lives

Since these are core insights, look for ways to help them emerge during the group discussion, as appropriate.

Using this activity for teaching in a business context

This activity works well in presentations of The Natural Step System Conditions for business contexts, such as with business school students or in workshops for the employees of a given company. For both of these contexts, it has several important features.

First, the consideration of everyday activities can be much less threatening than dissecting a business's products and inputs. The seemingly mundane examples provided in the accompanying activity sheets are effective at engaging people intellectually without making them feel on the spot, which might lead them to shut down emotionally and stop learning altogether.

Second, the activity provides an excellent springboard for business case studies. The numbers of such case studies are growing quickly, and they provide excellent inspirational examples for any business group. Since virtually any example can be helpfully described in terms of the four System Conditions, the activity is an effective low-stakes warm-up.

Conclusion

Consider the challenge of building a sustainable world as a balancing act between technical solutions and human solutions. On the one hand, we need more efficient technologies for transportation, agriculture, manufacturing and other activities that meet our needs. On the other hand, achieving sustainability will require changes in behaviours and attitudes.

Similarly, each individual has a cognitive and an emotional side, and both must be engaged for the individual to engage with the idea of sustainability. This activity takes an appealing and increasingly widespread cognitive framework, the four System Conditions of The Natural Step, and provides an activity that helps people to engage with the framework emotionally and intellectually.

One must both *understand* the connections and the big issues, and *feel* one's connection to the big issues. It isn't enough to *know* that recycling and energy efficiency matter; individuals often have to experience—even if just for an hour during a workshop—a sense of the big picture of which they are a part and which they can affect. This activity can be successful in providing that experience.

Appendix: mock-ups of activity sheets

Each activity sheet has the one-sentence action on one side and the list of related facts on the other. Each sheet is folded and stapled shut so the action is on the outside, while the facts remain hidden until the appointed time.

As part of the workshops that Good Company runs, I have used the following one-sentence actions on the activity sheets:

- 'I took out the garbage.'
- 'I drank a latté.'
- 'I drove my car to work.'
- 'I ate a hamburger.'
- 'I bought a T-shirt.'
- 'I got some new shoes.'
- 'I bought a computer.'

The full handout descriptions follow. (Note: The first action is followed by a box that summarises the four System Conditions. I recommend leaving this on each sheet for easy reference, but I have left it off the six other handout examples here in order to save space.)

First handout
I took out the garbage.

On other side of handout, stapled shut
Facts about solid waste:

- More than half (61.9%) of our waste stream is composed of paper (38.1%), yard waste (13.4%) and food waste (10.4%), which could all be recycled or composted (see Fig. 7.1). In the United States, we recycle about 28% of our municipal solid waste

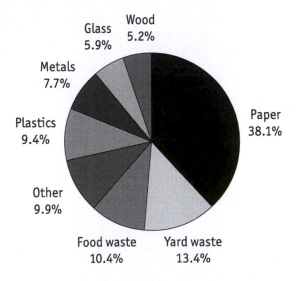

FIGURE 7.1 **1996 total waste generation: 209.7 million tons (before recycling)**

- The United States recycles 42% of all paper, 52% of major appliances and 55% of aluminium beverage cans

- Americans generate more waste every year, growing from 340 million tons of non-hazardous waste in 1997, to 390 million tons in 1999—a 15% increase in just two years

- Landfills have liners to prevent the escape of 'leachate', the toxic sludge that leaches out of the landfills. According to the EPA (Environmental Protection Agency), even the best liner and leachate collection systems will ultimately fail due to natural deterioration; 82% of surveyed landfill cells had leaks

- For every one job waste disposal creates, recycling creates 5–10 jobs. It is estimated that we spend US$100/ton to dispose of 'municipal waste'

- Hazardous waste landfill and municipal solid waste landfill appear to be similar in their ability to produce toxic gases. In a recent study, more than two-thirds (67%) of landfills tested produced one or more of the following persistent toxic gases: vinyl chloride, benzene, ethylene dibromide, methylene chloride, perchloroethylene, carbon tetrachloride, trichloroethylene and chloroform

- Landfills are a significant source of methane, a major greenhouse gas

System Condition 1: Nature is not systematically subject to increasing concentrations of substances extracted from the Earth's crust

System Condition 2: Nature is not systematically subject to increasing concentrations of human-made substances

System Condition 3: Nature is not systematically subject to degradation, destruction and over-harvesting

System Condition 4: We must systematically use resources efficiently and meet basic human needs

On open side of handout

I drank a latté.

On other side of handout, stapled shut

Facts about coffee:

- Coffee production (overwhelmingly for export) supports 20–25 million farmers and agricultural workers worldwide. Coffee exported to the United States comes primarily from South and Central America. (Coffee is the second most traded good after oil, measured by the total value of international trade)

- Coffee production traditionally has consisted of 'mixed-shade' systems in which coffee is cultivated together with trees grown for their fruit or wood (for timber or fuel). These mixed-shade systems support high biodiversity—more than any other agricultural system, and nearly as much as some natural tropical forest

- In the past 20 years, production has moved increasingly towards 'sun coffee', intensive monocultural systems that use high-yielding varieties and large amounts of fertiliser and pesticides. These more industrial systems typically have higher short-run yields, but they support very little biodiversity (and often interfere with the routes of migratory birds) and they result in much greater soil erosion and water pollution

- Fertilisers are usually produced from petroleum

- Farmers in developing countries often lack the equipment and training to use fertilisers and pesticides safely. Still, these chemicals are widely available and typically loosely regulated or simply unregulated

On open side of handout

I drove my car to work.

On other side of handout, stapled shut

Facts about cars:

- A typical car contains about 2,000 pounds (900 kg) of iron and steel, 250 pounds (110 kg) of plastic, 190 pounds (90 kg) of aluminium, 135 pounds (60 kg) of rubber and nearly 100 pounds (45 kg) of copper and other metals

- Approximately one-fourth to one-third of urban space is dedicated to cars in one form or another

- More than two-thirds of US petroleum consumption is used for transportation, which accounts for about one-third of US carbon dioxide emissions. Motor gasoline accounts for nearly two-thirds of transportation sector energy consumption

- About 45% of the steel used in cars is recycled scrap

- A steel mill uses about 31 gallons (117 litres; 250 pounds) of water to make a single pound of the steel that goes into the car

- Painting a car often entails a series of steps something like this:
    - Dipping in detergent and phosphate bath to clean
    - Dipping in zinc phosphate and chromic acid to prevent corrosion (zinc and chromium are both mined)
    - Four 'baking' stages, using huge amounts of energy and emitting VOCs (volatile organic compounds)

- Six additional coats: a sealant, an anti-chip layer, a primer, a colour coat, a clear coat and a noise reducing tar. The overspray becomes a sludge (containing polyvinyl chloride and solvents which are persistent synthetic petroleum-based chemicals), which gets shipped to a landfill

On open side of handout

I ate a hamburger.

On other side of handout, stapled shut

Facts about beef:

- A quarter-pound of beef requires 616 gallons (2,332 litres) of water to produce. Forty per cent of US beef cattle are fattened by mining ancient water from the dwindling Ogallala Aquifer of Colorado, Kansas, Nebraska and Texas

- More than 70% of US grain production is used to feed cattle and other livestock. Half of US cropland is planted with crops to feed animals

- Atrazine, a persistent chemical, is the most heavily used pesticide in US agriculture, including conventional grain production. It is the second most commonly detected pesticide in a national survey of drinking water wells

- The typical mouthful of American food travels 1,200 miles (1,900 km) from farm to consumer

- Chronic malnutrition (including protein deficiency) affects about 800 million people worldwide. Animal products (including beef) are an important source of protein for most people

- Most of the energy used in beef production is in the production of fertiliser for feed crops: a chemical plant heats methane (from natural gas) to form hydrogen, which is then compressed and superheated and added to nitrogen (from air), ending up with ammonia which is used as a fertiliser. The US economy consumes nearly one pound (0.5 kg) of ammonia per person per day, mostly as fertiliser

On open side of handout

I bought a T-shirt.

On other side of handout, stapled shut

Facts about T-shirts:

- Cotton is grown on 3% of the world's cultivated land and accounts for 25% of total global pesticide use

- Polyester is made from petroleum. (Only about 3% of petroleum in the US is used to make petrochemicals such as polyester; the remainder is used for gasoline and diesel fuel)

- Common textile dyes are regulated by the EPA as hazardous substances

- In 1989, there were 100 acres (40 hectares) of organic cotton cultivation in the United States. By 1994, there were 18,000 acres (7,300 ha). (That's still far less than 1% of all land growing cotton in the US)

- The US manufactures a great deal of cotton fabric. Much of this fabric is exported to Central America and the Caribbean, where it is cut and sewed into final products for re-export to the US

- Drying a T-shirt requires fully one-tenth as much energy as manufacturing it in the first place

On open side of handout

I got some new shoes.

On other side of handout, stapled shut

Facts about shoes:

- The United States imported US$26 billion worth of footwear in 1998, 59% of it from China and much of the rest from elsewhere in East and South-East Asia

- The synthetic materials in footwear are made from petroleum-based chemicals

- According to surveys, American women own 15–25 pairs of shoes and men own 6–10

- Leather is often treated with 'chrome tanning', a process involving chrome, calcium hydroxide and other strong chemicals

- Shoe soles are often made of styrene-butadiene rubber, a synthetic petroleum product. (Farmers in the tropics still harvest natural rubber from rubber trees, but two-thirds of the world's rubber is synthetic)

- Most athletic shoes are made by hand with solvent-based glues and luminous dyes, often quite toxic

- New Balance manufactures some of its shoes in the United States, where it pays its workers US$10 per hour plus health benefits. About 4% of the retail price of the shoes is labour; for shoes produced in Asia, labour costs are around 1–2%

On open side of handout

I bought a new computer.

On other side of handout, stapled shut

Facts about computers:

- A typical desktop computer requires a power supply of 150 watts, and computers take about 5% of electricity used in American offices. (In comparison, lighting uses 20–25%)

- The United States owns about 40% of the world's 300 million computers

- Manufacturing a typical computer generates about 139 pounds (63 kg) of total waste (including 49 pounds [22 kg] of toxic waste), 7,300 gallons (27,600 litres) of waste water and 2,300 kilowatt hours of electricity. Much of the toxic waste consists of VOCs and persistent organic pollutants (POPs)

- Circuit boards are plated with layers of copper and tin–lead solder. Tin is still mined in various countries around the world. Recycled lead (mainly from dead car batteries) meets 60% of US demand annually

- Mining and producing metals accounts for about 7% of global energy consumption

- Japan, many European countries and several Canadian provinces have implemented extended producer responsibility (EPR) laws and regulations for computers (and many other goods, from televisions and refrigerators to cars). EPR laws make companies responsible for products at the end of product life, thus encouraging them to design for re-use, recycling and/or disposal

Part III
Case studies

8

Easter Island
A CASE STUDY IN NON-SUSTAINABILITY[*]

David K. Foot
University of Toronto, Canada

The world is a lonely planet in the universe. while it depends on the Sun's gravitational force for its position and light/heat energy, its inhabitants rely entirely on its own earthly resources for their survival. These inhabitants, both human and non-human, are isolated. To date there has been no confirmed emigration from Earth and no visitations of species from elsewhere in the universe to planet Earth.

Easter Island is a lonely island on Earth. While it is subject to earthly weather and ocean patterns, it is remotely located in the South Pacific Ocean 3,200 km west of the nearest continent of South America. The first human inhabitants arrived from Polynesia about AD 400 and were isolated from other humans for over a millennium. They had to depend entirely on their own island resources for their survival.

Their legacy of giant stone statues, known as *moai*, suggests that, in spite of their isolation, the island population thrived and developed over this period much as it has on planet Earth. However, by the time that the modern explorers discovered Easter Island in the 18th century, something had gone terribly wrong with the society and its development. Most of the *moai* were smashed and toppled, and the people were headed towards civil war.

What went wrong on Easter Island? Recent research suggests that the demise of an advanced culture on the island can be attributed to the disappearance of one crucial resource on the island: wood (Flenley 1993). Wood is a renewable resource, but like all renewable resources it takes time for the resource to renew. Population growth and

[*] The author acknowledges with thanks the comments of Dr Georgia Lee of the Easter Island Foundation on an earlier draft of this chapter.

human decisions can result in insufficient time for effective renewal to take place. In this case, environmental collapse can precipitate economic and social collapse. This is what happened on Easter Island (Flenley 1993, 2001).

The recently documented history of Easter Island provides students today with a simple, easy-to-understand historical case study of an isolated society growing and developing but not practising sustainability through ignorance, neglect, self-interest or simply bad luck.[1] A careful review of this case study provides important lessons for the isolated inhabitants of planet Earth today. It also points to the ominous consequences of not practising sustainable behaviour for both individuals and society.

Geography, anthropology and history

Easter Island, also known by the indigenous name of Rapa Nui, is one of the most isolated places on Earth that is inhabited. A triangular-shaped island anchored at its corners by extinct volcanoes, Easter Island covers only about 117 km² (64 square miles) of area. Its maximum length is 24 km (15 miles) and maximum width 12 km (7.5 miles). It is located in the South Pacific Ocean 3,200 km (2,200 miles) west of Chile, the South American country to which it belongs. It is 2,000 km (1,250 miles) from the nearest inhabited land of Pitcairn Island, which has a population of under 50 people. The nearest inhabited islands further to the west are the Mangarévas which are 2,500 km (1,550 miles) away and the Marquesas which are 3,200 km (2,200 miles) away.

The island's latitude of 27° south (just south of the Tropic of Capricorn) provides a mild climate while its volcanic origins provided a reasonably fertile soil. The pollen record obtained from core samples from the volcano cauldrons using carbon dating methods shows that the island originally supported a great palm forest (Flenley 1993). Surrounded by an abundant ocean and supporting two extinct volcanic craters that contained natural lakes for fresh water, Easter Island appeared to have many attractive features for human habitation. There was, however, no sheltered anchorage for boats.

It is now widely accepted that the first inhabitants were a small group of seafaring Polynesians accidentally blown off course or deliberately in search of a new settlement, perhaps seeking a new kingdom or escaping warfare or overpopulation elsewhere (Bahn and Flenley 1992).[2] Archaeological evidence suggests that they arrived with a full panoply of colonisation goods. Popular perceptions are that the original settlement came in two large double canoes with up to 40 people, along with their sweet potatoes, chickens and other food items. This concept fits nicely with an orthodox opinion of continued Polynesian exploration and settlement across the vast Pacific Ocean that began out of South-East Asia about 1,000 BC or even earlier (Ponting 1993). After Tonga and Samoa, they then settled the Marquesas around AD 300 or earlier and then moved southeast to Easter Island and north to Hawaii in the 5th century and on to the Society Islands about AD 600 and New Zealand about AD 800 (Ponting 1993).

---

1 Alternative theories about the possible reasons for this behaviour will be examined briefly later in the chapter.

2 Bahn and Flenley (1992) provide a history of Easter Island. See also Fischer 1993 for additional information.

The basic social unit was the Polynesian extended family, which jointly owned and cultivated its land. Closely related households formed clans each of which developed its own centre for family and ceremonial activities. These settlements were scattered across the island in small clusters of huts around a ceremonial centre with crops grown in the surrounding fields.

As the population grew the settlements proliferated. Each was characterised by a large stone platform known as an *ahu* situated at the centre of the settlement near the coast (Bahn and Flenley 1992). Similar to those found elsewhere in Polynesia, these *ahu* were used for burials and ancestor worship. On these *ahu*, the Easter Islanders added statues quarried in volcanic stone from sites on the island. The main Rano Raraku quarry site contained basalt stone used for the statues and the second Puna Pau quarry site, about 8 km (5 miles) away, contained red stone used for topknots that were placed on top of the heads of the statues. White coral collected from the ocean was used for the eyes of the statues. Early statues were quite small, but as the society developed the statues were made larger. The largest statues on site were over 6½ m (20 feet) high and weighed over 80 tons.

By the 16th century hundreds of *ahu* had been constructed and over 600 of the huge statues erected. The population peaked in mid-century at around 10,000 (±3,000) people and then suddenly collapsed. Many partially completed statues and topknots were left in and around the quarry sites. Several statues still *in situ* in the Rano Raraku quarry measure up to 22 (65 feet) and weigh up to 270 tons.

Probably the most riveting mysteries of human history are those posed by various so-called lost or vanished civilisations, such as the Maya, Khmer and Anasazi peoples. Easter Island is no different. The first European to visit Easter Island was the Dutch admiral Jacob Roggeveen on Easter Sunday (5 April) 1722, which is why the island was given its European name (Bahn and Flenley 1992). Roggeveen found an island denuded of trees and inhabited by a society of about 3,000 people living primarily on agricultural produce. A Spanish expedition in 1770 reported a primitive society that lived in squalid reed huts or caves engaged in warfare and resorting to cannibalism in a desperate attempt to supplement the meagre food supplies available on the island (Ponting 1993). Subsequent visits by better-known captains such as Englishman James Cook in 1774 and Frenchman J.F. La Pérouse in 1786 confirmed previous reports (although the French considered the population 'happy'). They provided even lower population estimates[3] and noted that a number of the statues had been toppled and the platforms damaged.

The 'mystery' of Easter Island that emerged from these visits and lasted well into the second half of the 20th century was how a late stone age culture represented by the physically small, squalid and apparently barbarous population observed by the Europeans could have been responsible for creating, moving and erecting the massive stat-

---

3   The population continued to decline. In 1862 the Peruvians raided the island and enslaved many Rapanui who were sold as household servants or agricultural workers on mainland Chile. Subsequent epidemics (especially smallpox and tuberculosis) and missionary-induced emigration resulted in further population decline until only about 100 people remained by the mid-1870s. Sheep ranching began around 1870 and Chile officially annexed the island in 1888. Between 1888 and 1952 the island was used as a sheep ranch, which further denuded the landscape. The population gradually increased over this period. About 3,000 people live on Easter Island today.

ues that clearly represented the past glory of the culture. Was this an example of another lost civilisation?

The 'mystery' explained

The Europeans reported that the islanders believed that the statues had walked to the platforms under the influence of a spiritual power. The first and perhaps most well-known modern theory was advanced by Norwegian archaeologist Thor Heyerdahl (1950, 1958) who argued that the island was first settled by native South Americans from whom they inherited a tradition of monumental sculpture and stone work (similar to the magnificent Inca achievements). He attempted to demonstrate the feasibility of his theory by designing a reed raft (named the *Kon Tiki*) according to the specifications of early vessels and sailing it from South America to Easter Island. His raft drifted to the north and landed on one of the Tuamotu islands. Heyerdahl accounted for the decline of the population by the subsequent displacement of the South American culture by a less advanced Polynesian culture that resulted in a series of civil wars between the so-called 'long ears' and the 'short ears', which destroyed the complex society and reduced the population on the island.

Other more exotic theories have also been proposed to account for the mystery. An earlier Atlantis-type theory was proposed by John Macmillan Brown (1924) who argued that Easter Island is a remnant of a great continent (sometimes called Mu) that housed an advanced civilisation, which sank beneath the ocean. A more recent and even more exotic theory was proposed by Swiss writer Eric von Daniken (1970) who suggested that the Easter Island statues are part of the evidence (along with such unexplained phenomena as the giant Nazca Lines of Ica, Peru) of visitations by an extraterrestrial civilisation to planet Earth.

The accumulation of archaeological evidence, particularly over the last quarter of the 20th century, has resulted in an alternative explanation focusing on environmental collapse. Some authors postulate that planetary environmental changes may have precipitated the decline of Easter Island society (Hunter-Anderson 1998), while others argue that the islanders themselves were responsible for the environmental disaster that befell them (Bahn and Flenley 1992; Flenley 1993, 2001; Mieth *et al.* 2002; Diamond 2004).[4] These could be mutually reinforcing explanations.

The easiest explanation for a population and hence societal crash is that a climate change acted as the trigger if not the bullet for an environmental disaster. Hunter-Anderson (1998) argues that deforestation on Easter Island was the result of long-term climate change that started well before colonisation. This theory supports Orliac and Orliac's (1998) observation that Polynesians were well aware of their environmental dependence and enriched ecosystems rather than impoverished them. MacIntyre

---

4   Forest fires emanating from lightning strikes are another possible explanation for the destruction of the forest, but evidence of widespread fire has not been uncovered. Mieth *et al.* (2002) report localised burning for agricultural purposes on the Poike Peninsula dated to the 14th century. Also there is the possibility that volcanic activity resulting in earthquakes or tidal waves (tsunamis) knocked down the statues, but this would not explain the population and environmental collapse.

(1999) also cautions that the possibility that climate variability contributed to the collapse of Rapa Nui should not be overlooked.[5] He notes that relatively small decadal climatic fluctuations may mean that a marginally supportable population in one decade may become far too many a decade later.[6]

MacIntyre (1999) introduces a simple population growth model incorporating a carrying capacity that he calls the maximum supportable population (MSP). If this MSP were constant, the model would predict slower population growth with larger populations until a sustainable population was achieved after which no further growth would occur (or births would equal deaths). To explain the population crash on Easter Island, MacIntyre reduces the MSP starting from around AD 1350 to 1400. This decrease in the MSP can reflect the impact of global climate change or the effects of deforestation, or both. Once carrying capacity drops below the actual population society is in trouble. Note that in this approach population does not 'overshoot' the MSP, which is the usual explanation of environmental collapse (Meadows *et al.* 1992; Brander and Taylor 1998); rather, it is the MSP that drops below the population.

The more popular theory is consistent with this approach but attributes the decline in the MSP to deforestation perpetuated by the islanders themselves. Bahn and Flenley (1992) refer to this behaviour as cultural suicide whereby the islanders exploited their crucial forest eco-base so heavily that they did not give it time to renew and thus destroyed it. Subsequent soil erosion together with intentional fires to clear the land for more intensive agricultural use also contributed to further eco-base degradation (Mieth *et al.* 2002). The ensuing resource and food scarcity precipitated civil war, societal breakdown (including perhaps cannibalism) and a 70% reduction of the population over a relatively short period of time (roughly between 1550 and 1650). Other authors (Ponting 1993; Bush 1997) have adopted and popularised this theory.

As previously noted, the pollen analysis shows that, prior to colonisation, the island supported a great subtropical palm forest (Flenley 1993). The Easter Island palm, now absent from the island, was related to the still-surviving Chilean wine palm which can grow 25 m (75 feet) tall and 2 m (6 feet) in diameter. The tall branchless trunks of this palm provided ideal wood for constructing large canoes and for transporting and erecting large statues. It might also have been a valuable food source since its Chilean relative produces edible nuts and sap from which the Chileans make sugar, syrup, honey and wines.

The trees also provided a haven and nesting place for birds, which supplemented the islanders' food and contributed to the propagation of the palms through their droppings. The islanders also used the wood and palm leaves for housing and for cooking and heating. Finally, the forests were cleared to make land available for agricultural uses including the cultivation of sweet potatoes, taro and other food crops. Therefore, wood was a crucial and pervasive eco-resource supporting the islanders' food (fish, birds, nuts), economy (housing, cooking, heating) and culture (statues). The collapse

---

5   A variant on this theme is a prolonged drought as the cause of the environmental disaster, but as Flenley (1993) notes it seems unlikely that the forest should survive for 35,000 years, including the major climatic fluctuations of the last ice age and the postglacial climatic peak, only to succumb to drought once people arrived on the island.

6   However, after an extensive review of climatic information, MacIntyre (2001) concludes that the position of Rapa Nui appears to isolate it from major sources of interannual climate variability.

of this resource attributable to deforestation to satisfy the myriad of societal needs including agriculture resulted in the collapse of the society (Flenley 1993).

These findings are supported by core samples from the islanders' garbage dumps. Nearly one-third of the discarded food bones from the growth period (AD 900–1300) come from porpoises. The common dolphin, which is the Easter Island porpoise, weighs up to 75 kg (165 lbs) and generally lives out to sea. It could not have been hunted from the shore but must have been caught well offshore using seaworthy craft built from the large palm trees. Porpoise bones disappear completely from the dumps by around AD 1500, which suggests that the islanders had lost the ability to catch large fish by that time (Diamond 1997).

In addition to porpoise and an occasional seal, early islanders feasted on abundant sea and land birds that used the palms and other vegetation for food and nesting. Bird stew might have been seasoned with Polynesian rat meat, animals brought by the original settlers either deliberately or inadvertently.[7] Unlike the rest of Polynesia, rat bones outnumber fish (as distinct from porpoise) bones at Easter Island archaeological sites (Diamond 1997).

Recent pollen and garbage dump analysis together paint a picture of an island[8] that prospered and then perished due to the non-sustainable use of the society's crucial eco-resource, wood. Deforestation was well under way by AD 800, only four centuries after human settlement. Charcoal from wood fires becomes more abundant over time (indicating growing numbers of people), while the pollen of palms and other trees and woody shrubs decreases and then disappears. The pollen of grasses that replaced the forest simultaneously becomes more abundant. Sometime after AD 1400 the palm tree becomes extinct on the island, not only because of human use but also because the birds that dispersed its seeds died out and the now ubiquitous rats prevented its regeneration by eating the seeds.[9] The 15th century marked the end of Easter Island's palm forest (Bahn and Flenley 1992).

The destruction of the birds followed that of the forest. Every species of native land bird (owls, herons, parrots) became extinct. Colonies of more than half the seabird species were eliminated. In place of native birds, the islanders turned to domesticated chickens for food. Later on, human bones appear in the garbage dumps. Cannibalism is part of the oral history of the islanders, but there is a lack of verifiable evidence (such as slash marks on the bones) for cannibalism. By this time fires for cooking and heating were fuelled by grass, sedge and sugarcane scraps and increasingly took place in the underground volcanic caves where the islanders sheltered.

After a millennium of growth and prosperity (AD 400–1400), Easter Island reached a peak in population and culture in the early 1500s. While their statues were getting larger, their trees were getting smaller and ultimately disappeared. Without wood (and rope) to transport statues their culture suffered. Without wood to build canoes their porpoise and fish catches declined. Without trees the birds disappeared and the resulting soil erosion reduced their crop yields. Chicken production and perhaps cannibalism only replaced part of the lost food. People were starving.

---

7   Flenley (1993) notes that the mouse-sized Polynesian rat was regularly introduced by Polynesian voyagers wherever they settled as a source of protein food.

8   This elaboration closely follows Diamond 1997 and is consistent with Bahn and Flenley 1992 and Flenley 1993.

9   Mieth *et al.* (2002) provide additional evidence of rat-chewed palm nuts.

Local anarchy replaced social stability and a warrior class took over from the hereditary chiefs and cultural leaders in the 1600s and 1700s (Bahn and Flenley 1992). Spears and daggers appeared and people moved into caves for protection from both the elements and their enemies. By the mid-1700s rival clans were toppling and smashing each other's statues.[10] By 1864 the last statue had been demolished. This was the society observed and reported on by the early European visitors.

The non-sustainable use and ultimate disappearance of one crucial renewable resource, wood, ultimately led to the demise of the population and society. This is a familiar story of growing populations facing shrinking resources, especially renewable resources such as fresh water and arable land. Homer-Dixon (1994, 1999) documents recent examples where renewable resource scarcity has resulted in starvation and civil war.

Reasons for the demise

Why did this disaster happen? One explanation is ignorance. Even though they were isolated, the islanders had developed technology that surpassed most other societies of their era.[11] Certainly, they were not technically ignorant. The numerous *ahu* with their elaborately carved statues throughout the island are evidence of an advanced culture. Clearly they were not culturally ignorant. However, if people are not aware of their impact on their environment, then environmental degradation can result especially in growing populations. Polynesian peoples have a long history of being environmentally aware (Orliac and Orliac 1998), so this seems an unlikely explanation on Easter Island. Since the island is not large and each clan apparently had its own land, islanders were probably aware of the diminishing number and size of trees on the island even over their relatively short lifetimes. Also, it is very likely that there would have been a transferable oral tradition that would have indicated increasing difficulties in harvesting wood for its many uses by each succeeding generation.

A second explanation is neglect by individuals, by leaders or by society. When change is gradual and is measured in lifetimes or generations, it is easy to be neglectful. By the time that the change becomes rapid it is often too late for corrective action. On Easter Island the trees gradually became fewer, smaller and, as society adapted, maybe less important. The last palm disappeared around the mid-1400s, yet the real crisis did not appear until a century later by which time it was too late to do anything about the loss of the forest.

Economists frequently focus on the roles played by incentives and self-interest in human decisions. Unlike most Polynesian palms, the palm that was indigenous to Easter Island was a very slow-growing tree. The Jubaea palm normally requires 40–60 years before it reaches the fruit-growing stage and it can take even longer. Brander and Taylor (1998) estimate that the typical lifespan for islanders that survived infancy was

---

10   See footnote 4 for a possible alternative explanation.
11   Debate continues to this day on how the massive statues were transported and erected, and it is still not known how the heavy stone topknots were placed on the statues' heads (van Tilburg 1995).

around 30 years, so the length of time to tree maturity would have exceeded the life-times of virtually all islanders. Consequently, a programme of replanting and caring for seedlings would never have been of direct benefit to the cultivators. It would only ben-efit their children or grandchildren. Under these conditions the incentives for forest preservation, especially in a growing population, were not adequate. Moreover, even if recognised as a problem, these 'incentives' were determined by technical considera-tions (tree and human life expectancy) that would have been very difficult if not impos-sible to change.

Brander and Taylor (1998) apply a Ricardo-Malthus model of renewable resource use to Easter Island and demonstrate that overshooting and collapse is much more likely for slower-growing tree species than for the more usual faster-growing species. Moreover, they go on to point out that Easter Island did not present a favourable envi-ronment for efficient institutional adaptation. While environmentally aware, the islanders probably did not understand the biology of the forest–soil complex or the incentive effects of alternative institutional arrangements, such as population control or forest preservation. By the time that disaster was imminent and the population might have been mobilised into action, institutional change would have been too late. Because of their isolation, the planting of a faster-growing species was not an available option. In short, the islanders were confronted by extremely unfavourable technical incentives (palm maturity that outstripped human longevity and limited technical knowledge) and a limited number of feasible options. The pursuit of self or clan inter-est in this environment, particularly in response to the needs of a growing population, was understandable but led to their downfall.

Finally, both Orliac and Orliac (1998) and MacIntyre (1999) raise the possibility that Easter Island's population collapse could simply be attributable to bad luck. The possi-bility that climate played a significant role cannot be overlooked, although MacIntyre's (2001) extensive review concludes that climate variability was not a likely cause.[12] Nonetheless, a prolonged period of climatic conditions that negatively influenced palm growth or maturity could have been sufficient to tip a marginally supportable popula-tion into an excessive population and start the downward spiral even without popula-tion growth. Hunter-Anderson (1998) notes that deforestation had been occurring even before human habitation began. Even if not decisive, climate change could have been a contributing factor in Easter Island's demise, reinforcing the effects of deforestation by the islanders.

Conclusions

The lessons of Easter Island emerging from recent archaeological evidence are only too clear for the world today. A lonely island with no emigration valve for its growing pop-ulation and no ability to import knowledge or alternative technology from elsewhere, could not adapt to an environmental disaster that destroyed its economy and society. Non-sustainable harvesting of the crucial renewable resource, a slow-maturing palm

12   See also footnotes 3 and 4 for other examples of 'bad luck'.

that provided wood for canoes and food provisions (porpoises, fish, nuts, birds), as well as housing, cooking, heating and cultural activities (statue moving and erection), resulted in the collapse of a growing, wealthy and culturally advanced society, to become warfaring cave dwellers over a relatively short period of human history. The non-sustainable harvesting practices also contributed to soil erosion resulting in lower crop yields and the loss of other food sources (crops, rats), which could not be made up for using alternative domesticated foods (chickens, grass and, perhaps, humans).

Earth is a lonely planet with no emigration valve for its growing population and no ability to import knowledge or alternative technology from other parts of the universe. Its inhabitants too must survive on the once abundant resources within its realm. Fortunately, research that contributes to increasing knowledge and shared understanding of global environmental problems can lead to adaptation and solution of recognised problems. However, Earthlings must be careful not to be caught in the Easter Island trap of overshooting the maximum sustainable population or engaging in self-interested, non-sustainable behaviour that reduces the maximum sustainable capacity below the existing population. The recently documented history of Easter Island clearly demonstrates that, under these conditions, collapse of population, economy and society is inevitable.

Postscript

Easter Island has many advantages as a case study for teaching lessons in sustainability analysis:

1. It is now a documented historical example that clearly demonstrates the consequences of non-sustainable environmental practices. It does not present what might happen but rather presents what did happen

2. It is simple. It focuses on the non-sustainable use of a single resource (wood). Since wood is a renewable resource, the case study can avoid discussion of alternative substitutes associated with the sustainable use of non-renewable resources

3. The analysis can be easily extended to demonstrate environmental interconnectedness. Examining the implications of deforestation for alternative food supplies (porpoises, birds) and for soil erosion (and crop yields) quickly incorporates the richness of most environment analysis into this case study

4. It is an interdisciplinary case study. It integrates the natural and social sciences. Students with varied backgrounds can all contribute to the analysis. Anthropology, archaeology, biology, business, demography, economics, forestry, geology, history, mathematics, ornithology, political science, soil science, zoology and other disciplines can all be advantageously captured in this case study

5. Although not a business case study per se, it incorporates many of the concepts familiar to business analysis, such as incentives, constraints, diminish-

ing returns, property rights, information availability, cultural values and institutional structures, within a sustainability framework. It can also form the basis for sophisticated modelling and dynamic analysis, and an introduction to non-linear chaotic systems

6. It highlights the necessity for a comprehensive approach to developing a sustainability strategy. The case study demonstrates that high (technical and cultural) expertise is no guarantee of success with respect to sustainability. The existence of inappropriate incentives and unfavourable external forces (maturity cycles for trees and humans, and perhaps climate) played a decisive role in the outcome

7. It is likely that the islanders were all of a similar culture, so resource depletion and societal collapse cannot be attributed to multi-ethnic conflict, again simplifying the analysis

8. The isolation analogy with spaceship Earth is easy to grasp and the conclusions are unambiguous and easy to communicate. Non-sustainable environmental practices ultimately lead to economic and social collapse

9. It provides a framework for case studies of resource-dependent corporations, communities and countries that practise non-sustainable resource use, especially with respect to renewable resources

The sustainability lessons introduced and taught in this uncomplicated case study are numerous and pervasive. They clearly show the problem that a growing, reasonably well-informed, mono-ethnic, technically and culturally advanced society dependent on a single resource faces in ensuring that unregulated decisions that benefit individual members do not lead to collective ruin. This situation is what the world potentially faces today with climate change and other environmental sustainability challenges.

References

Bahn, P.G., and J.R. Flenley (1992) *Easter Island, Earth Island* (London: Thames & Hudson).

Brander, J.A., and M.S. Taylor (1998) 'The Simple Economics of Easter Island: A Ricardo-Malthus Model of Renewable Resource Use', *American Economic Review* 88.1: 119-38.

Brown, J.M. (1924) *The Riddle of the Pacific* (London: Fisher Unwin).

Bush, M.B. (1997) *Ecology of a Changing Planet* (Englewood Cliffs, NJ: Prentice Hall).

Diamond, J. (1997) 'Easter's End', *Doctor's Review* (from *Discover*), January 1997: 64-71, 139.

—— (2004) *Collapse: How Societies Choose to Fail or Succeed* (New York: Viking Penguin).

Fischer, S.R. (ed.) (1993) *Easter Island Studies* (Oxford, UK: Oxbow Monograph 32).

Flenley, J.R. (1993) 'The Palaeoecology of Easter Island, and its Ecological Disaster', in S.R. Fischer (ed.), *Easter Island Studies* (Oxford, UK: Oxbow Monograph 32): 27-45.

—— (2001) 'Forest and Civilization on Easter Island', in Y. Yasuda (ed.), *Forest and Civilizations* (New Delhi: Lustre): 55-62.

Heyerdahl, T. (1950) *The Kon-Tiki Expedition* (London: Allen & Unwin).

—— (1958) *Aku-Aku: The Secret of Easter Island* (New York: Rand-McNally).

Homer-Dixon, T.F. (1994) 'Environmental Scarcities and Violent Conflict: Evidence from Cases', *International Security* 19.1: 5-40.

—— (1999) *Environment, Scarcity and Violence* (Princeton, NJ: Princeton University Press).

Hunter-Anderson, R. (1998) 'Human vs Climatic Impacts, or Did the Rapanui Really Cut Down All Those Trees?', in C.M. Stephenson, G. Lee and F.J. Morin (eds.), *Easter Island in Pacific Context* (Los Osos, CA: Easter Island Foundation): 85-99.

MacIntyre, F. (1999), 'Is Humanity Suicidal? Are there Clues from Rapa Nui?', *Rapa Nui Journal* 13.2: 35-41.

—— (2001) 'ENSO, Climate Variability, and the Rapa Nui. Part II. Oceanography and Rapa Nui', *Rapa Nui Journal* 15.2: 83-94.

Meadows, D., L. Meadows and J. Randers (1992) *Beyond the Limits* (London: Earthscan).

Mieth, A., H.-R. Bork and I. Feeser (2002) 'Prehistoric and Recent Land Use Effects on Poike Peninsula, Easter Island (Rapa Nui)', *Rapa Nui Journal* 16.2: 89-95.

Orliac, C., and M. Orliac ((1998) 'The Disappearance of Easter Island's Forest: Over-Exploitation or Climatic Catastrophe?', in C.M. Stephenson, G. Lee and F.J. Morin (eds.), *Easter Island in Pacific Context* (Los Osos, CA: Easter Island Foundation): 129-34.

Ponting, C. (1993) *A Green History of the World: The Environment and the Collapse of Great Civilizations* (London: Penguin).

Van Tilburg, J.A. (1995) 'Moving the Moai: Transporting the Megaliths of Easter Island: How Did They Do It?', *Archaeology*, January/February 1995: 34-43.

Von Daniken, E. (1970) *Chariots of the Gods? Unsolved Mysteries of the Past* (New York: Putnam).

9

Suncor Energy
A COMPREHENSIVE APPROACH TO SUSTAINABILITY

Peter A. Stanwick and Sarah D. Stanwick

Auburn University, USA

In 1917, the Sun Company began operations in Canada. Over the past 85 years, the company has evolved into a fully integrated energy company now known as Suncor Energy. Over the past century, Suncor Energy has had to address a number of financial and environmental challenges to be able to survive in the highly competitive energy industry. Suncor Energy is renowned for its commitment to long-term environmental sustainability. Suncor's core purpose statement states in its *2001 Report on Sustainability* that its vision, in part, is to be 'a unique and sustainable energy company'.

In the 2001 report, *Stepping Forward: Corporate Sustainability Reporting in Canada* (Stratos Inc. 2001), Suncor Energy was recognised as the top firm in Canada addressing sustainability issues. In addition, Suncor Energy was one of the original companies to be included in the Dow Jones Sustainability Index.

Understanding Suncor Energy

Suncor Energy has a significant presence in the energy industry in Canada. It employs over 3,000 people and has over $10 billion in assets. A summary of Suncor's financial performance in Canadian dollars is shown in Table 9.1. In addition, Suncor yields, on average, over 200,000 barrels of oil or equivalent energy products daily. Suncor is com-

	2000	1999	1998
Production			
Natural gas (thousand barrels of oil equivalent/day)	20.0	22.6	24.7
Natural gas (thousand m³ of oil equivalent/day)	3.2	3.6	3.9
Crude oil and natural gas liquids (thousand barrels of oil equivalent/day)	121.1	119.0	109.9
Crude oil and natural gas liquids (thousand m³/day)	19.3	18.9	17.5
Refined product sales (thousand m³ refined product/day)	14.6	13.8	13.8
Financials ($ millions unless otherwise stated)			
Revenues	3,388	2,387	2,070
Operational earnings	414	167	175
Net earnings	377	186	178
Cash flow from operations	958	591	580
Return on capital employed (%)	16.6	8.3	9.5
Capital and exploration expenditures	1998	1350	936
Total assets	6,833	5,176	4,104
Year-end share price on Toronto Stock Exchange ($)	38.30	30.20	23.00

Values are in Canadian dollars

TABLE 9.1　Suncor Energy's economic performance indicators

Source: Suncor 2001a, 2001b

posed of three highly integrated business units which are: natural gas and renewable energy; Sunoco refining and marketing; and oil sands.

Suncor does exploration and natural gas development in the western part of Canada. The average daily production of natural gas is approximately 200 million cubic feet. Twenty per cent of the earnings are generated by Suncor's natural gas production. A summary of the Natural Gas division's financial performance is shown in Table 9.2.

In the province of Ontario, Suncor, operating under the Sunoco brand name, refines petroleum products and supplies gasoline to over 500 stations across the province. The refinery processes 70,000 barrels of crude oil daily. Sixteen per cent of the earnings generated by Suncor come from the Sunoco division. A summary of Sunoco's results are shown in Table 9.3.

	2001	2000	1999
Revenue	449	428	306
Production (thousands of boe/day)	33.4	40.5	51.1
Average sales price			
Natural gas ($/thousand cubic feet)	6.09	4.72	2.44
Natural gas liquids ($/barrel)	34.38	36.66	19.32
Crude oil ($/barrel)	33.92	29.50	20.94
Operational earnings	103	59	22
Net earnings	117	98	41
Cash flow provided from operations	280	238	172
Total assets	722	762	962

Values are in millions of Canadian dollars unless otherwise noted
boe = barrels of oil equivalent

TABLE 9.2 Summary of the Natural Gas Division's financial results
Source: Suncor 2001a

	2001	2000	1999
Revenue			
Refined product sales (thousands of cubic metres)	2,588	2,604	1,779
Sunoco retail gasoline	1,575	1,539	1,500
Total	*5,419*	*5,360*	*5,080*
Operational earnings	70	68	27
Net earnings (loss) breakdown			
Rack back	47	69	14
Rack forward	23	(1)	13
Others (tax adjustments)	10	13	–
Total	*80*	*81*	*27*
Cash flow provided from operations	165	174	103

Values are in millions of Canadian dollars unless otherwise noted

TABLE 9.3 A Summary of Sunoco's financial results
Source: Suncor 2001a

The oil sands deposits that are being developed by Suncor are located near Fort McMurray, Alberta. Suncor mines the oil sands and extracts the component bitumen from the deposits. The bitumen is refined and upgraded. Once upgraded, it yields refinery feedstock and diesel fuel. The average production is expected to be approximately 225,000 barrels of oil equivalent by the end of 2002. Almost two-thirds of Suncor's earnings (64%) is generated by the production of the oil sands. A summary of the Oil Sands division financial results are shown in Table 9.4.

	2001	2000	1999
Revenue	1,385	1,336	889
Production (thousands of bpd)	123.2	113.9	105.6
Average sales price ($/barrel)	29.17	31.67	23.84
Operational earnings	342	324	167
Net earnings	283	315	167
Cash flow provided from operations	486	655	405
Total assets	6,409	5,079	3,178

Values are in millions of Canadian dollars unless otherwise noted
bpd = barrels per day

TABLE 9.4 A summary of the Oil Sands financial results

Source: Suncor 2001a

Sustainability issues

As stated in its *2001 Report on Sustainability*, Suncor believes that being an integrated oil and gas company can create a parallel path with investing in alternative and renewable energy sources to generate increase shareholder value. As such, its commitment to environmental issues is extremely important. Suncor's environmental policy is shown in Box 9.1.

Suncor considers itself as a sustainable energy company which focuses on the interaction of three factors with sustainable development. As stated in its *2001 Report on Sustainability*, Suncor believes that sustainable development can be accomplished by understanding the role it plays in the relationship among a strong economy, social well-being and a healthy environment. This belief is based on the assumption that the ability to generate and supply energy that satisfies the needs of various stakeholders including customers, stockholders, employees, environmental groups, the government and local communities can yield positive benefits for all interested parties. Suncor identifies potential stakeholders by considering the potential impact Suncor operations may have with the interest group. Suncor evaluates who a stakeholder is based on: (1) how close the stakeholder is to the company or specific project (i.e. the impact on local

"At Suncor Energy, we care about the environment. We accept the responsibility entrusted to us to develop resources, conduct our operations and offer our products and services in a socially, economically and environmentally-responsible manner. To this end, we are committed to excellence in implementing standards of care than not only comply with legislated requirements but respond to the social, economic and environmental expectations of our communities, customers, shareholders, government and the public."

Box 9.1 Suncor's Environmental Policy

Source: www.suncor.com

communities); (2) the impact of their operations from a national and/or international perspective, the financial social and environment impact of Suncor's actions; and (3) whether Suncor's operations impact critical landmarks of public interest (i.e. environmental sensitive areas).

Suncor believes that there are a number of critical benefits to managing their stakeholders to help satisfy their needs. Suncor requests active involvement by the stakeholders in the decision-making process, leading to the resolution of issues by generating new ideas through alternative viewpoints. In addition, participation of the stakeholders allows information to flow to the interested stakeholders and creates additional shareholder value for Suncor.

Within its *2001 Report on Sustainability*, Suncor has developed a strategic framework to address the relationship between sustainability issues and the expectations of the stakeholders. The components of the framework are: (1) operational excellence; (2) stakeholder relations; (3) products and services for the future; (4) integrated decision-making; (5) public policy, education and awareness; and (6) organisational capabilities and commitment.

Suncor strives to obtain the maximum level of operational efficiency. This philosophy yields not only cost savings benefits but efficient operations which reduces the level of waste products to be disposed of. As stated previously, Suncor views all stakeholders as important in the development and implementation of strategic issues that pertain to sustainability. In addition, Suncor is committed to the development of alternative and renewable energy sources. The company also employs an integrative approach in its decision-making. As a result, tools such as the Life-Cycle Value Assessment system are used to include sustainability in the decision-making process. Suncor also develops numerous educational programmes for both employees and the public. It attempts to link its employees and various contractors to sustainability through its organisational capabilities and structure.

Examples of Suncor's involvement in sustainability issues are evident in a number of different areas. At its oil sands location in Fort McMurray, Suncor aided in the creation of the Cumulative Environmental Management Association (CEMA) which is responsible for accessing and monitoring the potential environmental impacts within the region. Suncor has raised the level of employment of native (aboriginal) people in the oil sands to 9% of the workforce in 2000, up from 3% in 1996. Suncor has adopted the 'ABCs of Climate Change' programme developed by the Energy Council of Canada. This is an outreach programme which demonstrates to employees and local residents how they can participate to lower levels of greenhouse gas emissions. Through the Suncor Energy Foundation, the company has donated approximately CAN$3 million to almost 300 non-profit organisations across Canada.

Managing sustainability issues at Suncor Energy

Being involved in three separate but related energy industries creates many challenges for Suncor. A summary of Suncor's environmental performance is shown in Table 9.5. One important area is air quality. While total production of oil products increased by 21% from 1996 to 2000, greenhouse emissions increased by only 12%. With Canada's committing to implementation of the Kyoto Agreement, Suncor must demonstrate its ability to reduce its level of greenhouse gas (GHG) emissions. As part of the Kyoto Agreement, Suncor, by 2010, must generate 6% less net GHG relative to its 1990 levels. Suncor views this as an opportunity and not a threat. As part of its sustainability strategy, Suncor has developed a range of actions in this regard, which include: management of GHG emissions; developing alternative energy sources; identifying new types of technology to reduce GHG levels; and implementing comprehensive training programmes for employees and educational programmes for the public.

Another environmental issue relating to air quality is the level of sulphur dioxide (SO_2) emissions emitted. SO_2 emissions are generated when hydrocarbons are burned in the oil sands to raise steam in the utilities boilers. In addition, SO_2 emissions are released as a by-product of the refining process and as an end result of the processing of 'sour' gas in the natural gas process. At Suncor, the level of SO_2 emissions has decreased by 70% from 1996 to 2000. In 2000, Suncor's daily production of SO_2 was 65 tonnes. Suncor is able to reduce the level of SO_2 emissions by a number of different actions. It implemented an acid scrubbing system based on limestone which captures the SO_2. In addition, it was able to capture sour gas that was previously burned or flared into the air and converted to an energy source. Another gas the emissions of which contribute to smog-like conditions is nitrogen oxide (NO_x). Suncor releases NO_x in the process of burning fuel for plant operations and diesel-powered mine equipment. The company has attempted to reduce NO_x emissions by working with manufacturers of heavy equipment engines and implementing low-NO_x burners at its plants through replacement or retrofit processes.

With six other international energy companies (BP Amoco, Chevron, Norsk Hydro, Statoil, Shell and Texaco), Suncor is investing in the CO_2 Capture Project. This project identifies and examines methods to capture CO_2 that is released by power generation stations and other sources. The current method used in this project is to store the captured CO_2 in underground caverns. CO_2 is currently injected in some oil fields to aid the recovery of oil.

	2000	1999	1998
Production			
Upstream processed volumes and gross production (thousand barrels of oil equivalent/day)	169.2	156.6	150.3
Upstream processed volumes and gross production (thousand m^3 of oil equivalent/day)	26.9	24.9	23.9
Downstream gross production (thousand m^3 refined product/day)	12.3	12.2	12.5
Total upstream and downstream production (thousand m^3 refined product/day)	39.2	37.1	36.4
Air emissions			
Greenhouse gas (GHG) (thousand tonnes CO_2 equivalent/yr)	6,366	6,169	6,160
GHG emission intensity (tonnes CO_2 equivalent/m^3 total production)	0.449	0.461	0.466
Greenhouse gas offsets (thousand tonnes CO_2 equivalent/yr)	200	0	0
Sulphur dioxide (tonnes/day)	65.0	78.7	69.5
Sulphur dioxide emissions intensity (kg/m^3 total production)	1.66	2.12	1.91
Nitrogen oxides (tonnes/day)	66.5	66.8	46.0
Nitrogen oxides emission intensity (kg/m^3 total production)	1.70	1.80	1.26
Benzene (tonnes/yr)	104.3	99.2	–
NPRI on-site releases to all media (tonnes/yr)	1436	932	1243
Energy consumption			
Energy use (million gigajoules)	80.4	74.9	74.0
Energy intensity (gigajoules/m^3 total production)	5.60	5.54	5.57
Water use			
Surface water withdrawal (million m^3)	74.0	73.6	69.1
Waste management			
Hazardous/dangerous wastes generated (tonnes)	8,910	7,143	–
Non-hazardous/non-dangerous waste generated (tonnes)	110,000	102,000	–
Waste recycled/re-used/recovered (tonnes)	5,073	5,264	–

TABLE 9.5 **Suncor energy environmental performance indicators** (continued opposite)

Source: Suncor 2001a

	2000	1999	1998
Products and services			
Ethanol blended into gasoline (thousand m³)	191.9	138.1	102.3
Sulphur content of gasoline (ppm)	192	209	301
Compliance			
Major incidents	4	1	7
Regulatory contraventions	120	209	182
Water quality discharge	1	2	-
Spills to watercourses	2	2	2
Regulatory fines ($)	750	20,000	1,000
EHS management			
EHS professionals on staff	52	45	37
Environmental capital expenditures ($ millions)	5	5	14

TABLE 9.5   (from previous page)

Source: Suncor 2001a

Alternative and renewable energy sources

Suncor has embraced the concept of green power as it pertains to sustainability. 'Green power' is the generation of electricity through alternative and renewable energy sources such as wind, solar, geothermal and biomass. In 2000, Suncor pledged to invest $100 million over the subsequent five years to identify and develop alternative and renewable energy sources.

Suncor has developed comprehensive criteria in determining which of those projects to invest in that could yield alternative and/or renewable energy sources. The criteria include: (1) not being a traditional energy source such as oil natural gas or coal; (2) the ability to reduce GHG emissions; (3) evaluation as an efficient energy conversion project based on a Life-Cycle Value Assessment; (4) having a minimal environmental impact; (5) having a positive social impact and producing a positive stockholder return.

Using these criteria in 2001, Suncor partnered with Enbridge Inc. to develop a fully functional 11 MW wind power project in Gull Lake, Saskatchewan. The SunBridge project is projected to increase by 10% the amount of energy generated by wind in Canada. As part of the evaluation of the environmental impact, SunBridge examines the impact of the project on the natural environment of the region, including birds and plant life. It is estimated that the wind power project will reduce levels of carbon dioxide by more than 30,000 tons per year.

Suncor is also examining the use of landfills as an alternative energy source and is attempting to capture methane and carbon dioxide gases released from landfills. Not only would this reduce the levels of emissions into the air; but these gases could be used as an alternative energy source. This project is incorporated within the overall plan of using biomass to generate fuels and convert solid waste into alternative energy by capturing released methane.

Accountability of sustainability issues

Suncor's Board of Directors are ultimately responsible for the development and implementation of the company's sustainability strategy. Operationally, Suncor relies on members of the senior management leadership team to actually implement and verify the actions of the company from a sustainability perspective. Accountability for sustainability-based action is further delegated to the business managers who are responsible for the day-to-day execution of the sustainability strategy. The business managers produce quarterly and yearly reports for the leadership team and the Board of Directors.

To evaluate the level of sustainability and assign accountability to the decision-making process the Life-Cycle Value Assessment (LCVA) system is used, as described in Suncor's 2001 progress report for *Canada's Climate Change Voluntary Challenge and Registry Program*. LCVA is a tool for evaluating the potential financial, environmental and social impact of the actions at Suncor. LCVA incorporates all the various stages of the life of a product or process, which could include the extraction of raw materials, the steps involved in the manufacturing process, the distribution process for the products, the use of the products by the consumer and the disposal or recycling of the products. With the LCVA tool, Suncor can identify areas in its operations that can be eliminated, minimised, or it can reduce the level of environmental impact. The cradle-to-grave identification process of LCVA allows Suncor to develop life-cycle thinking that incorporates life-cycle measurements in the decision-making process and supports sustainability strategic actions. Suncor uses the LCVA processes to link the actions of its operations and the environmental performance indicators. In addition, LCVA is incorporated into Suncor's continuous improvement programme which is used to increase efficiency within its operations. Furthermore, the use of the LCVA process allows Suncor to present a positive environmental image both nationally and internationally. Additional benefits of using LCVA are: it allows Suncor to reduce its risk of unexpected adjustments in environmental impacts; it improves the company's ability to implement a more comprehensive financial analysis by using full cost–benefit analysis; it improves its ability to accumulate current information pertaining to the environmental impacts of its day-to-day operations; and increases its ability to make more informed purchasing and technology adoption decisions.

Auditing of sustainability issues

Suncor's internal auditors evaluate not only the company's financial operations but also its environmental actions. Environment, health and safety (EHS) audits at a business-unit level are conducted every two or three years and an overall EHS assessment is done every three to five years. Areas that are included in the environmental audit are: (1) the level of GHG emissions; (2) the level of GHG intensity; (3) the level of SO$_2$ emissions; (4) the level of SO$_2$ intensity; (5) the number of major environmental incidents; (6) the number of regulatory contraventions; and (7) the amount of environmental penalties and fines. The environmental auditing procedures include: (1) accumulating data and evidence of the activities of all components of Suncor's operations; (2) conducting a comprehensive assessment of the internal controls used by Suncor to ensure environmental compliance; (3) the use of substantive testing, which gauges the overall reliability of the data presented; (4) providing an assessment by the auditors of the materiality of the errors made by Suncor in relation to the implementation of its sustainability commitment; (5) presenting to stockholders the results of the audit in the management letter which describes the audit and highlights recommendations made by the auditors.

References

Stratos Inc. (2001b) *Stepping Forward: Corporate Sustainability Reporting in Canada* (Ottawa: Stratos Inc. in collaboration with Alan Willis & Associates and SustainAbility, November 2001; www.stratos-sts.com/sts_files/stratos.full.report.pdf).

Suncor Energy (2001a) *2001 Report on Sustainability: Our Journey toward Sustainable Development* (Calgary, AL: Suncor Energy; www.suncor.com).

—— (2001b) *2001 Annual Report* (Calgary, AL: Suncor Energy; www.suncor.com).

—— (2001c) *Progress Report for 'Canada's Climate Change Voluntary Challenge and Registry Program'* (Calgary, AL: Suncor Energy; www.suncor.com).

10

Perspectives on accounting and society

TEACHING ACCOUNTANTS CORPORATE SOCIAL RESPONSIBILITY (CSR): A CASE STUDY

Kumba Jallow

De Montfort University, UK

Rapid changes in the business world should be reflected in developments in undergraduate university courses. This is often seen where the changes are of a mainstream nature—moves away from traditional manufacturing to a knowledge-based economy are reflected in the teaching to students of business subjects where, for instance, the implications of technological developments are incorporated into traditional management courses. However, the changes taking place in the latter part of the last century have been radical in their political and social implications, and yet it seems that teaching in universities has been slower to catch up with developments in the 'real' world. Pressure from managers and practitioners to respond to these changes is reaching universities (Hailey 1998), and courses are being developed to incorporate such interdisciplinary concepts as **sustainable development** and **corporate social responsibility**.

This case study investigates how one such course—a final-year undergraduate module—has been developed and implemented. It takes as its core the notion of accounting and accountability, and is delivered by accounting lecturers to business students following a range of business programmes. The course has been running for several years and has been amended as experiences are assimilated, and as knowledge is built up.

Background

Undergraduate Business Studies and Accountancy students usually progress through their studies in a fairly conventional way. They learn the tools and techniques of subjects such as marketing, accounting, strategy and human resources, and they are encouraged to apply these to real-world situations. They often graduate with a clear understanding of organisational structure and behaviour, and the mechanisms with which these can be promoted. However, what is often missing is the introduction of new paradigms with which to examine current business practice. In particular, courses that question the 'business as usual' approach and attempt to provide a framework that allows this investigation are often marginal to the educational experience of many business undergraduates.

Nevertheless, there are modules emerging that do seek to present a review of current developments in the greening of business. We see examples of this where an accounting course would include the techniques for accounting for environmental impacts, and explore the tools and mechanisms available in practice to enable this process. A more critical approach may be adopted, where the current systems are critically analysed to provide a means to make them more transparent. In this way students begin to understand why business activity may be creating greater environmental and social impacts, and what motivation there may be to continue with the 'business as usual' approach. It then becomes important to review the alternatives to current practice that may better overcome the barriers to improvement of corporate performance in this field.

Accounting and accountability

In order to bridge the gap between accounting as a technical subject and accounting as a social construct, it is useful to adopt a theoretical perspective that reflects the nature of accounting as a means of reflecting responsibility. Accountability is such a perspective:

> Accountability is 'the duty to provide an account (by no means necessarily a financial account) or reckoning of those actions for actions for which one is held responsible' (Gray *et al.* 1996).

There is an obvious link to the way in which accounting is traditionally taught—reporting being the end result of many accounting processes. Accountability is also a framework that sits easily within the wider framework of sustainability: how can we assess (or measure) our progress towards or away from a more sustainable world if we do not have an account of those actions that affect our journey?

Accountability therefore has a central position in the course on Corporate Social Responsibility as it allows the philosophies and theories to emerge that help to explain not only how organisations can become more responsible but also how society receives the information to judge their actions.

Course developments

In order to present alternative views of accounting and its effects, a module was designed which would place accounting in the context of its wider implications. The aims of the new module were:

- To allow business studies and accounting students to address the broader issues of corporate social responsibility (CSR) through the examination of accounting and reporting models that move towards greater transparency and accountability

- To develop a critique of current business practices to review the changes necessary to move towards sustainability

- To review the contribution that the accounting profession can make towards the greening of business

- To examine and critique current practice in corporate social reporting

The course is offered to final-year undergraduates who are taking accounting as a non-vocational subject. In other words, any student who has taken accounting subjects in the first and second year, but who is not following the programme resulting in an Accounting and Finance degree, is able to choose to take this course, or has to follow the course as part of their individual programme.

Details of the course content

Table 10.1 shows a typical list of lecture subjects. This shows how the theme of accountability is incorporated into the framework of sustainability, and how subjects build on the theme so that an incremental approach to CSR and sustainability is developed.

The course begins with an examination of some underlying philosophies. This may include a discussion of eco-socialism, eco-radicalism and eco-modernity. This gives scope to critique the role of business and attempt to impose a philosophical position in a normative sense, by discussing what position(s) should be adopted.

The course then moves on to look at significant business structures—multinational enterprises (MNEs). These are important because of their size, the amount of world resources they control, their political influence, and the part they play in our everyday lives. The MNEs' use of major non-renewable mineral and biotic resources can be associated with environmental impacts across their sphere of influence, which is global. This also includes human resources, and so issues of human rights, employee working conditions, and the effect on neighbourhoods and communities can be explored. This can then move on to a more practical investigation of the manner in which businesses can express their environmental concerns through their external environmental reports. There is also the opportunity to explore the accounting profession's activities

Week	Topic
1	Introduction: what the module is about, what is expected, assignments
2	Philosophical frameworks: eco-socialism, eco-modernity, eco-radicalism
3	Multinational enterprises and their role in environmental management
4	Current trends in environmental reporting: latest guidelines
5	Accounting for energy and waste: environmental accounting in practice. A case study
6	Management accounting in large organisations
7	Accounting in its social context
8	Social accounting and social audit book-keeping
9	Sustainability: issues in accounting
10	Gender issues in accounting and management
11	Free and fair trade
12	Summary and conclusions; role of the accountant in CSR

TABLE 10.1 **Perspectives on accounting and society: indicative lecture programme**

in this area, through initiatives such as ACCA's Environmental Reporting Award Scheme[1] and the Global Reporting Initiative (see Holland and Gibbon 2001).

The course moves within the boundaries of the organisation to examine the types of accounting processes that may help to make the business operations more transparent in respect of environmental impacts, and focuses on energy and waste. Here the course can examine examples from the world of business, through case studies of actual practice. There is also the opportunity to examine this through a critical lens.

The course also explores the social aspects of business activities. This follows the developments in social accounting, book-keeping and auditing. Two further subjects are included to stimulate and provoke student response: gender issues and fair trade. These subjects are specific examples of sustainability in action, as they examine the responsibility of business for issues beyond profit.

Students are thus introduced to the components of sustainability incrementally. The accountant's role can be established by examining the range of models available for 'accounting for sustainability'. Thus all the strands of the course are brought together.

By the end of the course, students have studied the accounting responses to sustainability, and should be able to place these in a management perspective. At the same time, students should be able to appreciate the multidisciplinary approach that is needed to understand the concept and the application of sustainability in business and in society.

1   See www.acca.co.uk.

Conclusion

The course has provided students with an understanding of the role of social responsibility in corporate life, and in particular the part that accountants and their profession may play in developing both the conceptual framework and the practical tools to enhance corporate social responsibility. The course is intellectually challenging and demands a great deal from the students—both in terms of participation and in self-directed learning. It does appear that those students who take the course begin to appreciate that the study of business and management involves concepts that push the boundaries beyond the traditional approaches, and will equip students with the abilities to become managers in the 21st century. Courses of this nature will continue to be required, and, if they are to be accepted into the mainstream, some of the barriers will need to be overcome. Nevertheless, there is both the interest and the will to allow these courses to continue to be developed within universities; their importance cannot be underestimated and must be recognised by those who provide resources for management education.

References

Gray, R., D. Owen and C. Adams (1996) *Accounting and Accountability: Changes and Challenges in Corporate Social and Environmental Reporting* (London: Prentice Hall).

—— and J. Bebbington (2001) *Accounting for the Environment* (London: Sage).

Hailey, J. (1998) 'Management Education for Sustainable Development', *Sustainable Development* 6.1.

Holland, L., and J. Gibbon (2001) 'Processes in Social and Ethical Accountability: External Reporting Mechanisms', in J. Andriof and M. McIntosh (eds.), *Perspectives on Corporate Citizenship* (Sheffield, UK: Greenleaf Publishing): 278-95.

11

The Anti-Junk Mail Kit

Chris Galea

St Francis Xavier University, Nova Scotia, Canada

John MacKinnon

Deloitte & Touche, Canada

On a beautiful autumn Friday in September 2005, two recent MBA graduates from Canada's best-known business school were having a lovely lunch out on the patio of one of their homes. A full bottle of Chianti had already been consumed and the decision facing these good friends was whether to open a second bottle and forgo any reasonable chance of doing any worthwhile work that afternoon. While discussing this possibility, the letter carrier arrived and deposited the usual stack of mail into the mailbox. 'Here we go again,' exclaimed, Enrique, 'another load of bloody junk mail . . . What would I pay to get rid of it!' 'Hmmm,' muttered his friend Jon, 'How about that second bottle? I believe I have an interesting idea.'

Junk mail in Canada

Flyers, coupons, free-distribution magazines, government mailings, catalogues and donation solicitations—four to five billion pieces of junk mail are distributed in Canada annually, weighing down the average Canadian household with 66 pounds of unsolicited mail per year. [1] Around 550 companies send out the bulk of this mostly unwanted mail.

1 *Calgary Herald*, 1 July 1997: A3.

The value of this junk-mail industry is substantial, with over $10 billion of revenue generated annually, and steadily growing at a rate of approximately 10% per year. A spin-off of the junk-mail industry is the business of the distribution of mailing lists. More than 2,000 mailing lists are brokered in Canada annually.

Canada Post

Canada Post is Canada's largest single distributor of junk mail. In 1991, Canada Post generated 30% of its revenues by distributing junk mail. Fully two out of every five pieces of mail delivered were unsolicited junk mail. In 1997, Canada Post made a drastic change and completely ended its delivery of 'low-end' junk mail, i.e. unaddressed bulk mail such as weekly flyers and coupons. Publicly, Canada Post appeared to be responding to environmentally fuelled public pressure; others saw the move as a strategic way of getting out of a highly competitive, low-margin business. Regardless of the reason, it was not an insignificant change. Ten thousand jobs and, reportedly, $30 million in annual profits were 'sacrificed'. However, Canada Post decided to continue to distribute 'premium' junk mail—much more lucrative mail that is targeted to specific households and postal codes.[2] Environmental sceptics noted that Canada Post has a state-granted monopoly to deliver all addressed mail in the country—hence its desire to hang on to this end of the junk-mail market.

Nevertheless, the move to end delivery of unaddressed junk mail and the resulting loss of jobs and revenue aroused intense political backlash. Unions and workers fought the decision to cease distribution of 'low-end' junk mail. Nevertheless, determined political pressure from environmental groups, an environmentally aware public and possibly some profit-focused self-interest convinced Canada Post to proceed with the termination of this service.

The message had been sent and heard: the public disliked junk mail. Nonetheless, as noted, Canada Post still carried 'premium' junk mail. Furthermore, although Canada Post remained the single largest distributor of junk mail, it was still responsible for only a third of junk mail distribution in Canada.[3]

The market for the Anti-Junk Mail Kit

Increasing public awareness of environmental issues in recent years has focused negative attention on the junk-mail industry. Most households view junk mail as an annoyance, and not as a service. The environmental impact of junk mail is well documented. In Canada, a minimum of two million trees are harvested each year for junk mail, 99% of which is thrown out or, at best, recycled. Millions of gallons of water and harmful

2   *Montreal Gazette*, 2 July 1997: A4.
3   *Calgary Herald*, 1 July 1997: A3.

chemicals are used to process the paper. Heavy metals are used in the inks.[4] From most perspectives, junk mail is nothing but a drain on the environment.

This environmentally informed dislike of junk mail was unarguable. So was the high nuisance factor with which the vast majority of the general public viewed junk mail. As a result, at least 99% of junk mail ended up in the trash. Thus, the question facing Jon and Enrique was not whether there was a consumer demand to stop junk mail but how big that demand was and how lucrative it might be. Thus was born the Anti-Junk Mail Kit: a service/product designed to eliminate most junk mail from those consumers who did not want to receive it.

The design of the Anti-Junk Mail Kit

When the two partners started to develop the Anti-Junk Mail Kit they had to keep a number of strategic factors in mind.

In 1997, Canada Post had agreed to not deliver junk mail to boxes with signs that read 'No Junk Mail!'[5] However, the policy was an all-or-nothing one: either you received all or none of the junk mail. Although not required to by law, other major distributors of junk mail in Canada likewise had agreed to honour the 'No Junk Mail!' signs. Canada Post did not, however, print or distribute these signs. Households were responsible for producing their own signs.

The all-or-nothing policy clearly did not work for everyone. Some people want some of the junk mail, such as coupons, free trial offers and announcements by non-profit organisations. For these people the problem is more complex. Some may not want any mail that does not bear an address. Others may want to be *removed* from specific mailing lists while some may want to be *added* to certain mailing lists. By law, a person must be removed from a mailing list on request. Unfortunately, many new mailing lists are generated each month, so a person who has requested to be removed from one list may easily end up on a new list.

Approximately 80% of organisations that compile mailing lists belong to the Canadian Direct Marketing Association (CDMA). The only way to resign from mailing lists is to submit regular requests to the CDMA. A more effective, albeit less practical, solution is to submit requests to the individual organisations.

For an Anti-Junk Mail Kit to be marketable, therefore, it must account for these issues. Using a bit of creative thinking and strategic planning, the two MBA graduates were able to address these factors and to design and develop their Anti-Junk Mail Kit.

---

4   www.stopjunk.com/environment.html
5   *The Globe and Mail*, 1 July 1997: A4.

The kit

Similar to the well-known environmental '3Rs', the Anti-Junk Mail Kit had six 'Rs', designed to help consumers to:

- Refuse unsolicited mail and flyers at their mailbox
- Remove their names from Canadian mailing and telephone lists
- Return junk mail back to its sender
- Restrict their names from being added to new junk mail lists
- Relate to Canada Post the refusal to accept further delivery of junk mail
- Request political action to stop junk mail at its source

The six major objectives of the Anti-Junk Mail Kit were to be achieved through various 'programmes' contained within the kit. Each is described in detail below.

The National Identification programme

This programme was created to fight unaddressed junk mail and flyers sent to households and businesses, and to pressure Canada Post and the federal government to stop junk mail at its source. The mailbox was the first line of defence. The sticker shown in Figure 11.1 was designed to be placed on mailboxes of the people who bought the kit.

FIGURE 11.1 Mailbox sticker

Three 'pressure' postcards (see Figs. 11.2–11.4) were also provided, to be sent to the local postmaster, to the President of Canada Post, and to the Cabinet Minister responsible for Canada Post, respectively. The goal of these cards was to create a wave of public pressure that would force political pressure to stop junk-mail deliveries.

These cards indicated that customers wanted junk mail halted at the mailbox and at its source. They also introduced the Anti-Junk Mail symbol, which was created to become synonymous with the kit through distribution and media coverage. The logo was affixed to each part of the kit to ensure the largest distribution possible.

Finally, the set of stickers shown in Figure 11.5 was included. Consumers were advised to return unsolicited mail by affixing these stickers to junk mail and returning it to Canada Post.

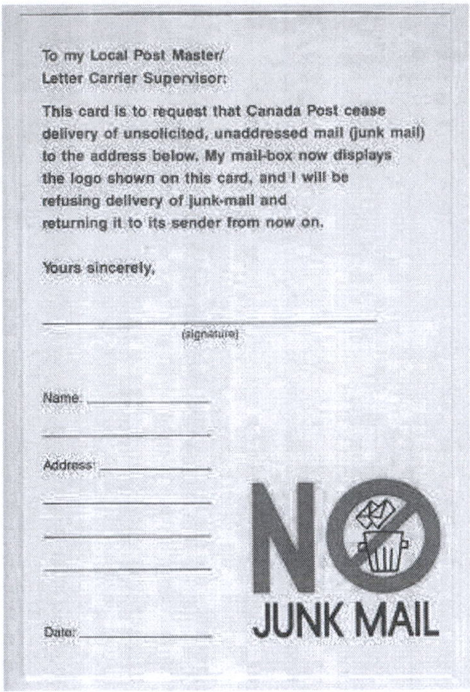

To my Local Post Master/
Letter Carrier Supervisor:

This card is to request that Canada Post cease
delivery of unsolicited, unaddressed mail (junk mail)
to the address below. My mail-box now displays
the logo shown on this card, and I will be
refusing delivery of junk-mail and
returning it to its sender from now on.

Yours sincerely,

_____
(signature)

Name: _____

_____

Address: _____

_____

_____

_____

Date: _____

NO JUNK MAIL

FIGURE 11.2 'Pressure' postcard to the
local postmaster

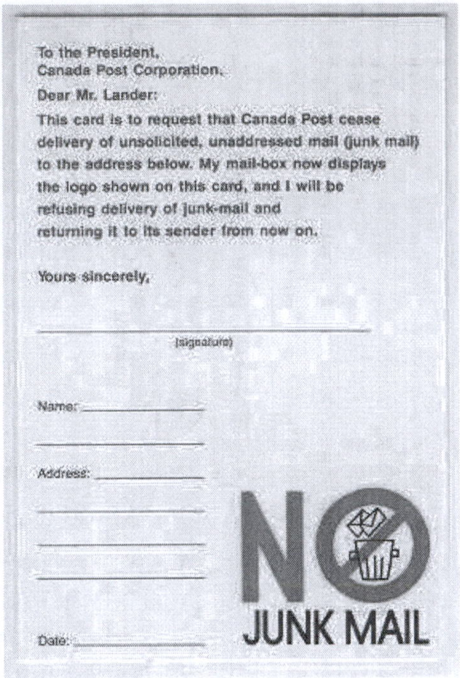

To the President,
Canada Post Corporation,

Dear Mr. Lander:

This card is to request that Canada Post cease
delivery of unsolicited, unaddressed mail (junk mail)
to the address below. My mail-box now displays
the logo shown on this card, and I will be
refusing delivery of junk-mail and
returning it to its sender from now on.

Yours sincerely,

_____
(signature)

Name: _____

_____

Address: _____

_____

_____

_____

Date: _____

NO JUNK MAIL

FIGURE 11.3 'Pressure' postcard to the
President of Canada Post

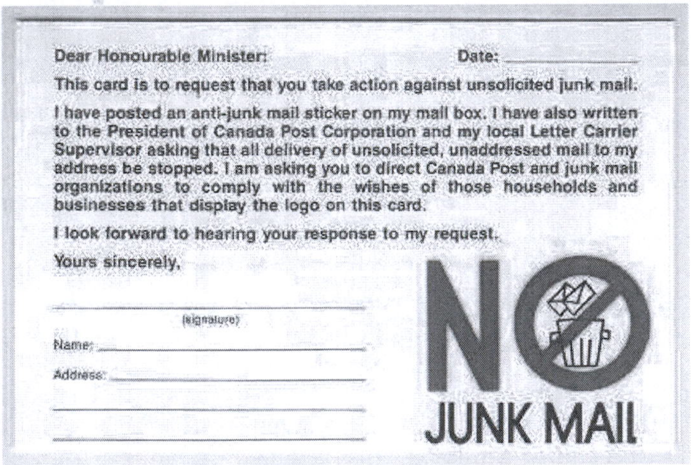

Dear Honourable Minister: Date: _____

This card is to request that you take action against unsolicited junk mail.

I have posted an anti-junk mail sticker on my mail box. I have also written
to the President of Canada Post Corporation and my local Letter Carrier
Supervisor asking that all delivery of unsolicited, unaddressed mail to my
address be stopped. I am asking you to direct Canada Post and junk mail
organizations to comply with the wishes of those households and
businesses that display the logo on this card.

I look forward to hearing your response to my request.

Yours sincerely,

_____
(signature)

Name: _____

Address: _____

_____

_____

NO JUNK MAIL

FIGURE 11.4 'Pressure' postcard to the Cabinet Minister responsible for Canada Post

FIGURE 11.5 Stickers to affix to junk mail

The strategy behind this part of the programme was quite Machiavellian. It was hoped that enough consumers would send return their junk mail back to Canada Post that it would create a massive bottleneck in their mailboxes. The resulting pressure would then force Canada Post to permanently deal with the issue.

Off-Mailing-List programme

This programme was designed to fight junk mail generated by mailing lists. As outlined earlier, approximately 80% of addressed junk mail is sent by over 500 organisations in Canada that compile or purchase mailing lists from various groups and list brokers. These organisations then conduct mass mailings either to solicit donations or to sell goods and services.

Most of these firms belong to the Canadian Direct Marketing Association (CDMA). The code of ethics for the CDMA obliges its members to comply with the following directive:

> Sec I. (iii) <u>All CDMA members must . . . delete the name of any consumer who has requested that he or she be removed from mailing and telemarketing lists.</u>

The CMDA also provides a centralised 'Mail Preference Service' which should remove people from lists on request. Unfortunately, new mailing lists are generated regularly and, within a few weeks, people's names end up on new mailing lists. Consequently, the only way to delete names off mailing lists permanently is to write regularly to the CDMA. Writing to each of the 500 members of the CDMA may be the most effective method, but it is highly impractical.

To address this impracticality, people who bought the Anti-Junk Mail Kit received the card shown in Figure 11.6 which authorised Mail Choice Inc. to request the deletion of their names from direct mailing lists. Mail Choice promised to contact the junk mail firms monthly. This card also provided for the option to receive specific types of junk mail (e.g. from charities or environmental groups), thus giving some consumers the option of receiving only the direct mail that they wanted.

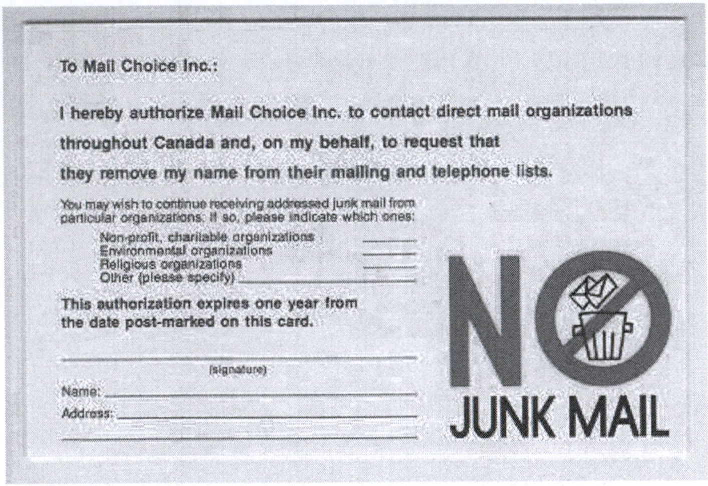

To Mail Choice Inc.:

I hereby authorize Mail Choice Inc. to contact direct mail organizations throughout Canada and, on my behalf, to request that they remove my name from their mailing and telephone lists.

You may wish to continue receiving addressed junk mail from particular organizations. If so, please indicate which ones:
Non-profit, charitable organizations _____
Environmental organizations _____
Religious organizations _____
Other (please specify) _____

This authorization expires one year from the date post-marked on this card.

_____ (signature)
Name: _____
Address: _____

NO JUNK MAIL

FIGURE 11.6 Authorisation card to Mail Choice Inc.

The second part of this programme was designed to fight the remaining 20% of junk mail not generated by mailing lists. Consumers were told to affix the stickers shown in Figure 11.7 to any remaining addressed junk mail that they still received after using the other procedures for removal from mailing lists. This system was set up because the sender typically has to pay for return postage. It was hoped that this would act as an incentive for the sender to stop sending unwanted junk mail.

FIGURE 11.7 Stickers to affix in order to return to sender

The Mailing List Prevention programme

This programme was designed to fight junk mail before it was generated. Customers were given three cards like the one shown in Figure 11.8 to send to organisations to which they subscribe or belong, and request that these organisations not sell their names to junk-mail organisations. The customer could choose what mail they wanted to be stopped.

Part 2 of the Mailing List Prevention programme contained stickers that customers could affix next to their signature whenever they provided their names and addresses

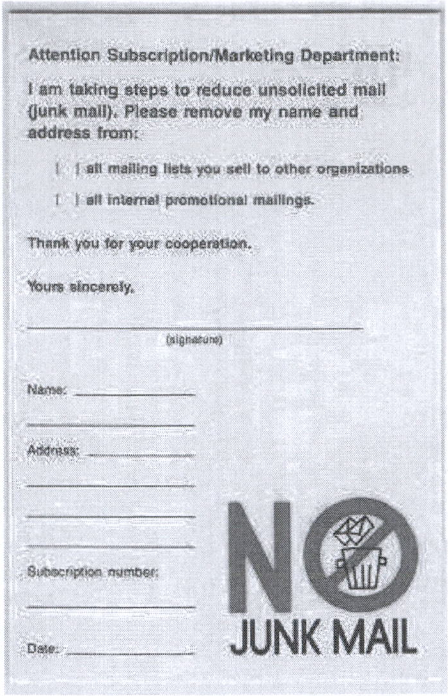

Attention Subscription/Marketing Department:

I am taking steps to reduce unsolicited mail (junk mail). Please remove my name and address from:

[ ] all mailing lists you sell to other organizations

[ ] all internal promotional mailings.

Thank you for your cooperation.

Yours sincerely,

______________________ (signature)

Name: ____________

Address: ____________

Subscription number: ____________

Date: ____________

NO JUNK MAIL

FIGURE 11.8 Card to request that names are not sold on

to an organisation (e.g. when subscribing to a magazine or filling out a warranty card) (see Fig. 11.9).

The Violators programme

Lastly was the Violators programme, a weapon of last resort for the Anti-Junk Mail Kit. It was designed to confront firms that refused to comply with customers' requests that they stop sending junk mail. Customers would send the card shown in Figure 11.10 to Mail Choice Inc., who would take further steps to remove customers' names from the mailing lists of the offending firms.

Environmental Impact Card

The illustration shown in Figure 11.11 was displayed prominently in the package to illustrate clearly just one environmental impact of the junk-mail industry. Furthermore, the paper portion of the kit was printed on recycled paper and displayed the recycling logo.

FIGURE 11.9 Anti-junk mail stickers to affix whenever completing a form

Packaging

Components of the kit were packaged in a clear zip-lock bag that measured 6" × 9" (14 cm × 22 cm) and which was sealed with the two-sided card displayed in Figure 11.12. Cards and stickers were designed and positioned in the package to allow for optimal display and ease of reading of the contents for potential buyers. The aim was to make the kit as self-explanatory as possible and to facilitate stand-alone retail sales. The inside of the two-sided card (Fig. 11.13) instructed consumers on how to use the kit. Note that the kit was perforated to allow for peg-boarding at checkout counters or at stand-alone impulse-buying displays.

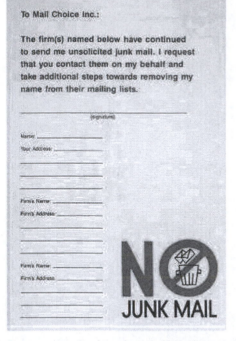

FIGURE 11.10 Card to alert Mail Choice to persistent violators

FIGURE 11.11 Card highlighting the environmental impacts of junk mail

FIGURE 11.12a The Anti-Junk Mail Kit as it appears on the shelves (front)

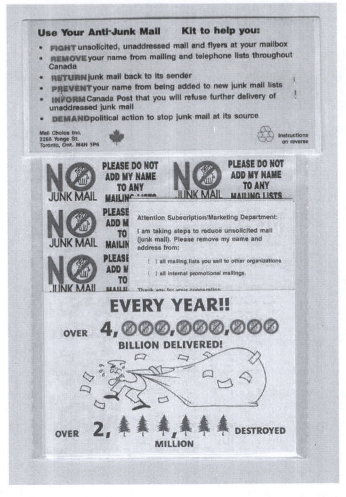

FIGURE 11.12b The Anti-Junk Mail Kit as it appears on the shelves (back)

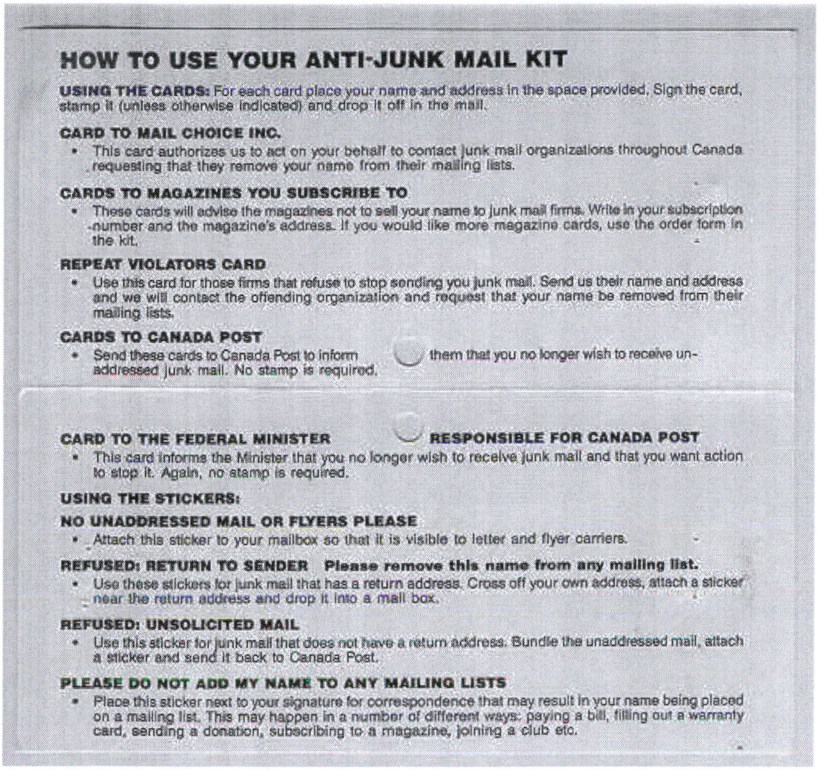

HOW TO USE YOUR ANTI-JUNK MAIL KIT

USING THE CARDS: For each card place your name and address in the space provided. Sign the card, stamp it (unless otherwise indicated) and drop it off in the mail.

CARD TO MAIL CHOICE INC.

- This card authorizes us to act on your behalf to contact junk mail organizations throughout Canada requesting that they remove your name from their mailing lists.

CARDS TO MAGAZINES YOU SUBSCRIBE TO

- These cards will advise the magazines not to sell your name to junk mail firms. Write in your subscription -number and the magazine's address. If you would like more magazine cards, use the order form in the kit.

REPEAT VIOLATORS CARD

- Use this card for those firms that refuse to stop sending you junk mail. Send us their name and address and we will contact the offending organization and request that your name be removed from their mailing lists.

CARDS TO CANADA POST

- Send these cards to Canada Post to inform them that you no longer wish to receive un-addressed junk mail. No stamp is required.

CARD TO THE FEDERAL MINISTER RESPONSIBLE FOR CANADA POST

- This card informs the Minister that you no longer wish to receive junk mail and that you want action to stop it. Again, no stamp is required.

USING THE STICKERS:

NO UNADDRESSED MAIL OR FLYERS PLEASE

- Attach this sticker to your mailbox so that it is visible to letter and flyer carriers.

REFUSED: RETURN TO SENDER   Please remove this name from any mailing list.

- Use these stickers for junk mail that has a return address. Cross off your own address, attach a sticker near the return address and drop it into a mail box.

REFUSED: UNSOLICITED MAIL

- Use this sticker for junk mail that does not have a return address. Bundle the unaddressed mail, attach a sticker and send it back to Canada Post.

PLEASE DO NOT ADD MY NAME TO ANY MAILING LISTS

- Place this sticker next to your signature for correspondence that may result in your name being placed on a mailing list. This may happen in a number of different ways: paying a bill, filling out a warranty card, sending a donation, subscribing to a magazine, joining a club etc.

FIGURE 11.13 Instructions

The next steps

Mail Choice Inc. believed that it had created a unique, environmentally beneficial and consumer-friendly product. Because the initial start-up costs were so low, the partners did not conduct any formal market research. Everyone they asked thought that the Anti-Junk Mail Kit was a great idea, and expressed personal support. People's desire to rid themselves of junk mail was clear and emphatic. Encouraged by the enthusiasm of everyone who saw the progress of the kit, the two partners decided to take the plunge. They approached various printing companies who agreed to print the various components of the kit. The initial print run was set at 25,000, giving a variable cost of less than $1 per kit. Working late into many nights the partners began to put together the 25,000 kits, all the while thinking about the best way of distributing them. At this stage they didn't have a formal plan but believed that they key to success was to conduct the speediest national roll-out possible. They thought this was necessary for the following reasons:

- Junk mail had high news value. Mail Choice Inc. wanted to take advantage of free media coverage generated by the saturation strategy

- The kit had could easily be copied. Mail Choice Inc. wanted to discourage competition before it began. The goal was to saturate all major markets before rival firms could begin

- The partners believed that effectiveness was linked to high participation rates across regions. Mail Choice Inc. wanted to pool feedback from these regions to increase the pressure on Canada Post, the junk-mail industry and the government to address the issues

- After a number of nights of lack of sleep, the partners were finally finished. They had 25,000 Anti-Junk Mail Kits packaged and ready for distribution.

'What next?' the two partners wondered?

12

The war of the woods
A FORESTRY GIANT SEEKS PEACE[*]

Monika I. Winn

Faculty of Business, University of Victoria, Canada

Charlene Zietsma

Richard Ivey School of Business, University of Western Ontario, Canada

'Bill, can I talk to you?' Tom Stephens, CEO of MacMillan Bloedel Ltd, the giant of British Columbia's forest industry, ushered Bill Cafferata, his Chief Forester, into his office.

'What's up, Tom?' Cafferata asked, trying to sound casual. Stephens had been CEO of MacMillan Bloedel (MB) (and Cafferata's boss) for exactly six months. During that time, he had replaced nearly the entire senior team with their former subordinates. Cafferata had survived the cleansing, and, after 26 years with the company, he was not especially anxious to end his career that day.

'I hear you've been having troubles with the Forest Project team. There is talk about some pretty loud arguments. My consultants tell me you guys were going to self-destruct, and that I'd better have a back-up plan in place. What do you think? Are you going to make it?'

Cafferata paused, knowing the next few words could give him enough rope to hang himself with. 'Yep, I think we're going to make it,' he said.

'Okay, go to it,' Stephens said.

A little relieved, a little uneasy, Cafferata went back to his office, determined to come up with a way to make the Forest Project work. The Forest Project was a cross-functional team of senior managers and experts that were charged with finding a new way

---

[*]   *Disclaimer:* The views expressed in this chapter are those of the authors alone, and do not necessarily reflect the views of the organisations mentioned.

	1997	1996	1995	1994	1993
Financial position					
Total assets	$4,559	$4,830	$5,271	$4,679	$4,397
Current liabilities	724	833	1,102	973	544
Long-term debt and liabilities	2,239	2,000	2,068	1,828	2,124
Shareholders' equity	1,596	1,997	2,101	1,878	1,729
Total liabilities and shareholders' equity	4,559	4,830	5,271	4,679	4,397
Results of operations					
Sales	$4,521	$4,267	$4,327	$3,781	$3,121
Costs and expenses	4,624	4,179	3,897	3,373	2,818
Operating earnings (loss)	(103)	88	430	408	303
Other income and expenses	(265)	(37)	(150)	(227)	(249)
Net earnings (loss)	(368)	51	280	181	54
Financial and statistical data					
Capital expenditures in continuous operations ($ millions)	90	220	503	232	167
Market price range High	21.70	20.10	21.62	23.50	23.62
Low	14.15	16.50	16.00	15.50	16.12
Earnings (loss) as percentage of sales—continuous operations	(3%)	2%	7%	6%	6%
Return on average common shareholders' equity	(22%)	2%	15%	11%	3%
Common shares outstanding (000s)	124,414	124,377	124,336	123,754	123,732
Number of employees (excludes discontinuous operations)	10,592	10,966	10,523	10,189	9,810

TABLE 12.1   MacMillan Bloedel five-year financial data (all dollar items, except market price range, are in Canadian $ millions)

Source: MB 1998 Annual Report

of managing forestry. Cafferata's first meeting with Stephens about the forest project one-and-a-half months earlier had been just as short. Stephens had conducted a 90-day strategic review of the company to determine what needed to be done to turn it around. Every aspect of the company had come under the scrutiny of Stephens and his 'Council', 'a team of high-priced, best-of-breed consultants', according to Cafferata. The management team from each division made a presentation to the council. They were told to make the business case for their units, mindful of Stephens's admonition: 'Sell it, milk it or grow it: get off your lazy assets.' Change came with a heavy dose of fear. The results of these presentations had become corporate legend: middle managers would come into work the day after their team presented to find that the team ranks had been decimated, and they were now in charge of making the 'revised' presentation. Any indication of MB's typical bureaucratic, divisionalised thinking was punishable by pink slip (i.e. termination of employment).

When Stephens took the helm in September 1997, MB had had six consecutive quarters of sizeable losses, and the losses were growing. Stephens was fond of saying 'MB is worth more dead than alive.' The company's market capitalisation was lower than the amount that could be generated by breaking it up and selling the parts. MB's five-year financial data are shown in Table 12.1. But Stephens felt that MB had other problems . . .

The 90-day review turned up several priority areas in need of major overhaul. One issue particularly critical for regaining financial success was how MB secured its supply of raw material, wood. As soon as the review was complete, Stephens had invited Cafferata, MB Chief Forester and VP, into his office.

'Bill, we need to find a new way to do forestry. I'd like you to form a team to find some solutions. I'll give you a budget of $1 million, and a time-frame of 90 days. Are you willing to take on this challenge?'

And so the Forest Project began.

The Forest Project team

Cafferata put together a cross-functional team comprising senior managers and experts in the areas of logging, silviculture,[1] conservation biology, forest ecology, forest growth and yield projections, and the social aspects of environmental affairs. The team included Cafferata himself, a professional forester. The team consulted outside experts from academia, government and industry research groups, along with internal experts.

Each team member brought a particular perspective to the table. Group members had gone to considerable lengths just to *understand* each other's issues, and disagreements had been fierce. In some ways, the diverse and often contradictory views in this group mirrored the myriad of conflicting views and pressures the company was exposed to externally. In line with corporate objectives for 'safety, respectability and outrageous success', the group agreed that they needed to find a way to:

---

1   Silviculture refers to the cultivation of trees; it aims to maximise long-term growth and yield of the forests through activities such as planting, thinning, fertilising and pruning.

- Increase conservation of old-growth forests

- Find a harvesting compromise to suit environmentalists and the public

- Achieve both of the above in ways that would enable the company to:
  - Protect employee safety first and foremost
  - Achieve forest certification (ISO 14001, FSC [Forest Stewardship Council], etc.) to ensure market access
  - Maintain employment at current levels
  - Meet or exceed regulatory obligations
  - Improve profitability

It was a wish list: no one really dreamed they would be able to meet all of the objectives, especially since many of them seemed directly in conflict and would involve trade-offs. However, this list of objectives gave each member of the group a set of targets to aim for. The team then decided that the best approach would be to split up into subcommittees, each of which was charged with becoming an expert in one of the key aspects of forestry that were currently at odds: biodiversity and conservation, harvest levels, yield and employment, safety, silviculture, social issues and profitability. The subcommittees had six weeks to do their research and generate recommendations for discussion by the whole team.

Company history, products and markets

MacMillan Bloedel had grown from a small British Columbia-based company at the start of the (20th) century, to one of the world's foremost forest products companies. The company managed 5 million acres (2 million ha) of productive timberlands (2.7 million acres [1.1 million ha] in British Columbia), which supplied most of its fibre requirements. Timber from these lands was cut in MB sawmills and further processed into lumber, panel boards, engineered lumber, containerboard and corrugated containers. Over decades of expansion, MB had entered and exited such businesses as pulp and paper, shipping, lumber distribution, and others, and had made investments and acquisitions in Europe, the US, Hong Kong, Australia and across Canada. By the late 1960s, MB had become Canada's 14th largest industrial corporation based on sales, and by the late 1970s employed 24,500 people in logging camps, sawmills and panel board plants, newsprint, pulp, fine-paper and paper bag plants in Canada, the US and the United Kingdom.

In 1996, MacMillan Bloedel had sales of CAN$5.043 billion, 13,497 employees, and harvested 5,716,000 cubic metres of logs, the equivalent of approximately 5.7 million telephone poles (1 m^3 = 35.315 cubic feet). The company consisted of three major business segments: building materials as its core business segment (67% of sales), paper (14%) and packaging (17%). Both paper and packaging were industries in overcapacity; MB's largest growth came from its building materials segment, particularly its value-added wood products. Figure 12.1 illustrates the products that come from the forest industry. MB had operations in each of these areas, and a research centre that devel-

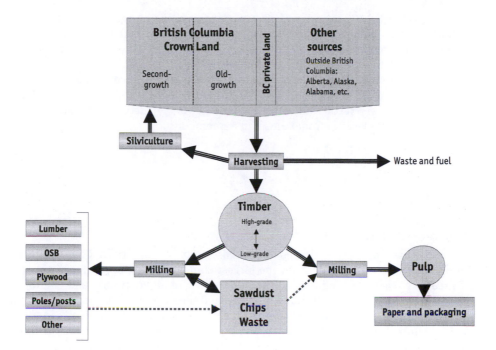

FIGURE 12.1 From trees to products: British Columbia forest industry value chain

oped new value-added wood products. MB also had a network of distribution centres across Canada and the US.

The biggest forest products company in British Columbia (BC), a Canadian province that depended heavily on forestry for both employment and revenue, MacMillan Bloedel had come under heavy fire from many sides. Environmentalists filled the media with negative publicity about MB's practice of clear-cut logging old-growth forests and organised boycotts on company products. The early symptoms of the Asian financial crisis gave management serious headaches: 40% of their lumber sales went to Asia. MB's biggest market was the US, but it could not easily sell its product there because the Softwood Lumber Agreement imposed restrictive quotas on Canadian lumber exports. Growing environmental concerns by customers in Europe, plus demands for environmental certification, limited access to that market. By the late 1980s and early 1990s, environmental controversies in coastal British Columbia had focused international media attention on the company's logging operations. Now, environmental groups had pressured several of MB's customers into cancelling purchasing contracts because of its logging practices. It seemed like the whole world was changing, but MB was out of step. Stephens claimed that MB was in danger of losing its 'social licence to operate' in BC.

Inside the company the situation was equally difficult. Devastating confrontations with unions had led to labour disruptions. A poor safety record did little to improve labour relations or cost efficiency. Low productivity, combined with the highest wages in the industry, placed MacMillan Bloedel at a big competitive disadvantage in global markets. And British Columbia, with its strong unions, high wages, complex forestry legislation, and significant environmental pressures already had a reputation for its 'unfriendly' business climate. Industry experts spoke of the 'BC discount', the lower stock prices (relative to earnings) accorded to BC logging companies.

Facing such tough times, MB executives talked about cutting costs and waiting for things to turn around. Investors, however, decided they would wait no longer and initiated a shareholders' revolt. CEO Robert B. Findlay was replaced by Tom Stephens, a seasoned corporate veteran from Denver, in September 1997.

The environment

Government regulation

For many years, forestry had been the largest employer in British Columbia and, through fees and tax revenues, was the largest contributor to the provincial government coffers. The forest industry accounted for 50% of all BC exports and employed 275,000 people in BC. The government owned and managed 95% of the forestland in BC (about 60 million hectares or 150 million acres; 1 hectare = 2.471 acres). In other countries, forests were usually privately held. In BC, however, the government granted forest companies long-term timber licences to cut and manage particular plots of land in return for stumpage fees.[2] Forest companies were required to adhere to harvesting, replanting and environmental regulations, which included the Forest Practices Code, a complex set of regulations claimed to mandate sustainable forest management practices (though environmentalists claimed otherwise), and the government mandated Land Resource Management Planning process, requiring public consultation with stakeholders prior to the approval of logging plans for a particular area.

Environmentalists

Starting in the late 1980s, local, regional and international environmental groups waged highly visible protests against forest management in British Columbia. Their major emphasis was on the practice of clear-cutting, which sometimes caused erosion of mountainsides and destroyed fish-bearing streams, habitat for wildlife (including bears and mountain lions), and high-biodiversity, old-growth rainforests. Two other major concerns were the cutting of old-growth forests (with trees ranging from 140 to over 1,000 years old) and the failure to preserve sufficient quantities of wilderness in parks. As the largest licence holder of public lands on the BC coast, MB was especially heavily targeted. Initially, MB ignored the environmentalists. But, when their voices became too loud, MB vigorously defended its practices of clear-cutting old growth. After all, it was largely in compliance with regulations governing logging in BC, and its prac-

---

2   Stumpage refers to a volume-based tax paid by BC forest companies for timber they harvest on public lands; it is a substantial source of revenue for the government.

tices were deemed by the forestry profession to be the best for the long-term growth and health of the forest, as well as the safest way to log.

Much of the steam behind the environmental movement in BC came from outside Canada. Money was raised in international campaigns (particularly in Europe), and large US foundations funded and participated in campaigns. A number of American actors, Robert Kennedy Jr and California state senator Tom Hayden travelled to Clay-oquot Sound on Vancouver Island's rainforest coast to support protests in the early 1990s. US Vice President Gore publicly pressured the Canadian government into converting public lands to parks and wilderness areas. A local logger's response to Robert Kennedy Jr's visit to BC typified the view of many BC citizens: 'I've worked hard for 44 years,' says Doug Pichette, 'and now I've got to listen to an outsider who had life given to him on a silver platter tell me what's wrong with my economy and government. Who does he think he is?'.[3] Environmental groups targeted US and European customers, threatening them with smear campaigns if they did not put public pressure on MB and other BC forest companies. BC observers considered it quite ironic that the US and Europe, where most of the forests had been cut years ago, were calling on Canadians to bear the full costs for conserving a 'world resource'.

The tourism industry and native land claims

An industry of growing importance, tourism, was also against clear-cutting: ugly bald clear-cut patches on mountainsides destroyed 'viewscapes' along the 'inside passage' to Alaska, the route travelled by a large number of cruise ships. Native land claims also created an element of uncertainty for BC forest companies. BC's aboriginal peoples claimed areas in excess of the total land area of the province (due to overlapping claims among different native groups). In the landmark 'Delgamuukw' decision, the Supreme Court of Canada ruled in December 1997 that aboriginal groups have rights in lands used or occupied by their ancestors, ranging from limited use up to aboriginal title. This decision called the security of long-term tenures on the lands licensed from the government into question. Until treaties were negotiated and court challenges settled, however, it was difficult to predict exactly what effect native land claims would have.

Changing global markets

BC forest companies exported 88.3% of their products (for revenues of approximately CAN\$13.24 billion) in 1998. The largest customer was the US, followed by Japan and Europe. The Softwood Lumber Agreement, negotiated between Canada and the US to run until 2001, restricted the volume of softwood lumber that could be exported into the US. Because MB had historically sold to the Asian market, a very small amount of US quota was allocated to it. When the 'Asian flu' (Asia's financial crisis) hit in 1997, MB turned its attention to 'greener' European markets, which had become increasingly competitive. MB had a very poor environmental image in Europe because of environmental groups' campaigns there. Already, two of MB's customers had cancelled their

3 *Globe & Mail*, 6 November 1993: A1.

contracts. And, now, a number of other customers were starting to ask MB staff what they were going to do to get the environmentalists off their back.

Environmentalists were calling for forest products to be independently certified as coming from sustainably managed forests. Three major certification schemes were emerging in BC forestry, each with different criteria. Environmentalists favoured the Forest Stewardship Council scheme, of which European furniture giant IKEA was a founding member. The Forest Stewardship Council had not yet developed criteria for certification of forestry in BC, but indications of what would be required could be identified based on criteria already established in other forest ecosystems.

Stephens sets new strategic objectives

When Stephens took the helm of MB in September 1997, the board of directors had given him a broad mandate for change. He refocused MB's strategy on its core building materials business and aggressively addressed the areas that hampered cost-competitiveness in its BC base. Under the new CEO, MB exited the paper business and medium-density fibreboard, closed its MB Research Center and shut down large numbers of distribution centres, sawmills, wood remanufacturing and packaging plants all over Canada and the United States. By late 1998, he had downsized MB's workforce to 9,000 from 13,500 employees worldwide, and he had completely restructured the senior management team.

During the restructuring, MB's formerly bureaucratic culture was to be shaken to its core: everyone in the company, managers and unionised employees, clerks, loggers and executives, would co-manage projects to turn the company around. Three key objectives were to guide every corporate action: to become the safest forest company, to become the most respected forest company, and to be outrageously successful. Given the company's recent dismal performance in each of these three areas, many observers thought they might as well be shooting for the moon.

The Forest Project subcommittees report their findings

In the spring of 1998, the Forest Project team met as a whole, for what was going to be a long day of heated debate. Six in-depth reports and memos summarising their recommendations had been distributed to all a few days earlier. It was clear that differences in perspectives had become more pronounced. Over the six weeks, each subcommittee had gained both a deeper understanding and a greater appreciation of their respective positions. The mood was tense. No one knew how to resolve what appeared to be unsolvable differences. The meeting started promptly at 8 a.m. With executive summaries of the six research reports in front of Team Leader Bill Cafferata and each team member, each of the six subcommittees presented its recommendations.

Executive summary

Task. Assess alternative logging methods' impact on current and future yield of fibre.

Clear-cutting. Any replacement of clear-cutting with a partial cutting method (variable retention) will increase cost significantly and lead to an immediate drop in harvest levels due to cutting fewer logs. Current methods allow us to cut an average of 8 cubic metres (m^3) of timber per year per hectare. To estimate the annual cut using variable retention (and ignoring the future declining yield effect), we subtract the percentage of retention from 8 m^3 per hectare, and multiply by the hectares planned to cut per year.

We also expect a future volume reduction. Using studies and simulation, we also expect the forest to regenerate more slowly as the trees left standing reduce the light to and growth of the 'understorey' tree crop. There are no studies or data for periods longer than 5 years on the effect of alternative silvicultural systems in coastal BC. The following are our best projections of yield reductions: at an average 25% retention rate, over 5 years: negligible; over 10 years: 2–3%; over 20 years: 5–6%.

Other considerations: Any variable retention system requires more and better roads, which have a very negative environmental effect. Also, clear-cutting mimics natural disturbances, providing a diversity of habitats for various species. Further, switching away from clear-cutting methods would cause a backlash from loggers, their unions, the industry and others in BC for giving in and selling out to environmentalist demands.

Old growth. About half of the 5.7 million acres in BC managed by MB currently contains old growth; the other half is second growth. However, second-growth forests are not ready to cut, requiring about 80-year cycles. Old growth will continue to account for the majority of MB's harvest for another 25 years. Many environmentalists are calling for a complete halt to the cutting of old-growth forests and to clear-cutting. If MB stopped both, harvest levels would be reduced to 20–25% of current harvest levels in the short term, to average about 2 million m^3 per year over 20 years. MB could not afford to maintain its operations at that harvest level, with implications not just for the solid wood business, but also for nearly every other business we are involved in.

Recommendation. In sum, our projections of yield reductions would hurt MB financially and would lead to lay-offs, plant shutdowns and devastation of the many forest dependent communities, all of which is clearly contrary to our landlord's, the government's, objectives for forest land use and social welfare. The harvest level subcommittee therefore recommends that it is critical to protect our access to fibre and that we must continue the use of clear-cutting old growth (except in designated preservation areas). If we continue to fight the environmentalist threat, eventually our customers will ignore environmentalists because they need secure access to wood.

Box 12.1 Technical report of the harvest level subcommittee to the forest project team

Executive summary

Task. Assess MB's role in retaining habitat conservation and biodiversity.

Biodiversity. Based on data from ecological and biological studies, large, contiguous blocks of old-growth forests are important for the maintenance of habitat and for plant and animal species biodiversity. Current parkland, though substantial, may not be of sufficient size or contiguity. Our estimates suggest that approximately 10% of MB's current old-growth holdings is in fragmented blocks of insufficient size to maintain biodiversity. Our old-growth strategy then should focus on cutting those fragmented blocks and retaining more of the wood in the large, contiguous blocks.

Wildlife habitat. Other areas provide valuable habitat for animal or plant species, but would not require the same level of preservation and high retention as the contiguous old-growth sections. These 'habitat zones' make up approximately 25% of our land base. The remaining 65% of the land base is prime timberland that could be subject to intensive forestry. These numbers are our best estimate.

Other values. Aesthetic and other values must also be considered, although results will depend on a social decision process and will likely be subject to significant controversy (e.g. road building carves up habitat; tourism requires access so that people can view wildlife). To determine exactly which blocks are to be preserved and to what extent, whether to conserve habitat, biodiversity or other values, we need to commission further scientific studies, and we need to develop a process that considers multiple stakeholders.

Recommendation. The biodiversity and conservation subcommittee concludes that, to meet market demands, it is essential for MB to show commitment to retaining old-growth forests; this should be done by distinguishing forest areas based on their primary value for biodiversity, for wildlife habitat, for primary logging and for aesthetic values.

Box 12.2 Technical report to the forest project team: biodiversity and conservation

Executive summary

Task. Review and assess MB's options to harvest, regenerate and grow forest crops.

Review. The two principal silvicultural systems available are clear-cut (removing all trees in an area) and selection (maintaining continuous, uneven-aged forest cover). To maximise the growth and yield of trees, MB has relied mostly on clear-cut (e.g. 93% from 1994 to 1996).

Variable retention. This selection system, developed to address a wide array of forest management goals, offers an alternative to clear-cut systems that focus exclusively on tree growth and yield. Variable retention follows nature's model by retaining part of the forest after harvesting. We know that dead standing trees, decaying wood on the forest floor, and diverse tree sizes and canopy levels are important as wildlife habitat. Variable retention retains these structural features, as habitat for a host of forest organisms, ensuring managed stands will be more similar to natural forests.

Different levels of retention can be used in different areas. Retention can be dispersed throughout a cutblock (individual trees or small groups) or aggregated (clumps or patches). For both safety and ecological reasons, aggregates are preferable. Clumps of trees are also more attractive to look at than a clear-cut site. Studies from other regions suggest that at 70% retention of ecologically sensitive areas, the effect on wildlife, biodiversity and aesthetics is negligible. At 40% retention, biodiversity may be affected, and aesthetics are obviously affected, but wildlife habitat is likely to be sufficient for most species. At 25% retention, habitat is affected, as are viewscapes. This level of retention may be best suited to areas that provide relatively poor habitat to begin with. At 10% retention, this method is not very distinguishable from clear-cutting.

Recommendation: A switch to variable retention will allow for other forest values such as habitat, aesthetic appearance and biodiversity. Variable retention is flexible enough to be adapted to specific terrain and it allows us to use high-value trees more effectively. While the annual harvest level would be somewhat lower, and trees do not grow as quickly or as plentifully under variable retention, we are convinced we could maintain forest productivity over the long term. Furthermore, we could obtain environmental kudos by being the first in BC to broadly implement this harvesting method.

Box 12.3 Technical report on silvicultural options to the forest project team

Executive summary

Task. Assess implications and feasibility of selection logging methods on worker safety.

Technical assessment. Using a partial harvest system (i.e. selection or variable retention logging) is more hazardous than clear-cut logging. Interlaced canopies of old-growth forests are littered with broken limbs and debris, and workers may be exposed to falling debris if they must work under or near a partial canopy. The risk can be reduced significantly through planning and training with deliberate attention to safety. An example is to develop harvest plans that allow loggers to fell trees into open areas; this is more easily achieved when trees retained on the site are clustered, instead of dispersed. Extensive retraining of loggers would be needed as well. With both, we might be able to log almost as safely using a partial harvesting system as we do now with clear-cutting.

Costs and risks. Safety is critical to MB, and we are already spending considerable resources to improve our safety record. Additional costs would be required to retrain loggers in a new method. There is also the risk that workers' compensation premiums would go up. We have worked with the BC Workers' Compensation Board to examine safety records of partial cutting systems elsewhere. The Board concedes that partial harvesting may be done safely and is willing to take a close look at a new system before deciding whether to increase our insurance rates upwards, if we adopt such a system.

For years, MB has been adamant that partial harvesting is hazardous and MB (and the industry) has argued that clear-cutting is the only safe way to log. It would be difficult to now try to convince our employees otherwise; it would be an especially hard sell to our unions. Logging safely can only be achieved with a new system if union members are willing to participate in training and to be flexible during the initial experimental stage.

Recommendation. The subcommittee concludes that selection logging can be done safely, only if we invest in the necessary planning and training, can convince the loggers of the new system, and can gain the trust of and work closely with the unions throughout the transition and learning phase.

Box 12.4 Technical report of the safety subcommittee to the forest project team

Executive summary

Task. Assess MB's key stakeholder groups and gauge their reaction to MB changing its logging practices to methods other than clear-cutting.

Overview. The report is based on focus groups, discussions and formal modelling methods of stakeholder groups to identify their needs, values and 'zones of tolerance' (the range of actions within which they are unlikely to complain about MB). We summarise perspectives of these key stakeholders: customers, environmentalists, competitors, government, unions, workers and communities, and the general public.

Customers. Many MB customers face significant pressure from environmental groups to publicly announce that they will no longer purchase products from old-growth forests and, particularly, forest products from MB. Examples are threats by Greenpeace to initiate consumer boycotts, media stunts targeting specific companies, and company email systems jammed with thousands of protest emails. Some customers have succumbed to the bullying tactics, cancelling contracts; others want MB to defuse the problem. Meanwhile, worldwide supplies of industrial wood fibre are abundant. BC is one of the world's largest producers of high-valued appearance-grade wood from old-growth forests, but, once customers find other sources, winning them back may be impossible. A good option for MB is to get certified using either ISO 14001 or Canadian Standards Association criteria for forest practices. MB may find itself with serious market access problems in the very near future, unless environmentalists' customer campaigns can be stopped.

Environmentalists. Local and international groups would likely welcome a proactive move by MB but, over the long term, might want no cutting in BC. Besides, environmentalists are generally distrustful of corporate interests and may interpret any proactive move cynically.

Government. Two ministries have jurisdiction over forestry. Owing to Ministry of Forests regulatory constraints, it is currently difficult to change silviculture systems: one, the full utilisation policy does not allow us to leave behind deadfalls and clumps of trees; and, two, requirements for average annual cuts over 5 years mean we could lose some of our timber land allocation if we depart from clear-cutting. If our alternatives can successfully deal with environmentalists' concerns, and they do not result in job losses, government officials may be flexible with respect to these regulations. They understand that logging of all old growth on the coast may be at risk if we don't make changes. The government, as owner of the majority of the forestlands managed by MB in BC, has the final say. MB must also continue to participate in the land resource management planning process, which covers five-year periods. The Ministry of the Environment supports a move to alternative silviculture methods that protect the biodiversity of plant and animal species better than does clear-cutting, so they are less likely to impede a change in practices.

Unions, Workers, Communities. The **unions** are concerned with maintaining logging and sawmilling jobs, the long-term future of the industry in BC, and worker safety. **Workers** themselves are also currently concerned with safety issues, feeling that their

Box 12.5 Research report to the forest project team: stakeholder management
(continued over)

safety is compromised by environmentalists (they have experienced nasty confrontations during blockades, and terrorist acts such as vandalism of equipment, tree spiking and the burning of a bridge). Moving to a variable retention system raises other issues. For years, the industry has asserted (in part to defend itself from environmentalists) that clear-cutting is the only safe way to log. Using another system may be difficult for employees to accept. Workers also fear that their jobs and communities will disappear if environmentalists succeed in converting logging lands to parks. Workers have also experienced being painted as 'the bad guys' by environmentalists, losing respect in their communities, and some saw their children harassed at school. In sum, we predict that workers and unions would accept a compromise that preserved current employment levels or reduced them by only up to 10%, while maintaining worker safety. The many remote forestry-dependent **communities** in BC lobby heavily for logging companies to continue to work in their areas and to provide high-paying jobs. Without regular forestry work, these towns would die, as little other work exists. Some communities attempt to diversify (e.g. into tourism) and to preserve the beauty of their towns, and would prefer to have trees left for aesthetic purposes in certain areas.

Competitors in BC. Fellow BC coastal forestry firms and our industry lobby group, the Forest Alliance, were mostly against changing to a selection harvest system. Such a decision would be 'caving in' to environmentalists' demands and giving away BC's economic prosperity, with new demands sure to follow. They favour continuing to resist environmental pressures. To date, MB has been the primary target, bearing the brunt of this fight, and others have few incentives to change. Some companies expressed support for new practices, recognising they could be the next targets, and that market access could become blocked for all BC logging firms.

The 'general public'. BC citizens have grown weary of the war of the woods, with cynical views about both companies and environmentalists: companies want only profit and are against conservation; environmentalists are extreme, and international groups are not sensitive to the economic dependence of BC on the forest industry. One result from focus groups was that people would trust only what environmentalists and forest industry people could agree on.

Recommendation. MB's objective is to become the most respected forest company. We recommend moving to variable retention in the most proactive way we can. We identified as potential benefits: rebuilding trust with the public and regaining a good reputation; reduced pressure from environmentalists on us and our customers; providing us with breathing room to develop a long-term strategy. We see no other option.

Box 12.5 (from previous page)

Executive summary

Task. Assess cost and profitability factors of switching to selective harvest methods.

Current conditions. The lumber market is currently in a down cycle and, after fees paid to the government, MB loses money on wood cut on public lands. Profits from private land are used to offset those losses. The reason behind this is that government regulations (designed to maintain stable employment levels for forestry and sawmill workers) require that annual cuts be maintained within certain limits. We expect government to be somewhat flexible with respect to changes in silvicultural and harvesting systems, but it is politically unlikely that large variations in employment levels would be tolerated. MB itself would have trouble retaining good workers if our cut varied significantly from year to year. Not all lumber prices are in down cycles at the same time, however (e.g. hemlock prices may be high when cedar prices are low). Currently, we have no choice but to cut all types of trees together, taking whatever price the market provides.

New scenarios. If we moved to a variable retention harvesting system, we could cut for value, not volume: we could focus on taking the trees out of the forest that have the highest value at a given time, providing we maintain forest biodiversity over time. We can maximise economic benefits by harvesting stands when their dominant species command the best price. To work, this approach requires a sufficient inventory of government-approved cutblocks—something we are currently unable to get. Focusing on positive margin stands could increase the average margin and net earnings of the company, but reduce the harvest, and leave trees for habitat, aesthetics and other non-timber values on sites not currently profitable to log.

Significant capital is required to transition to new harvesting systems. Large equipment currently used for clear-cutting requires large logging roads, and road building is very expensive and has a large ecological impact. If cut sites are to be significantly reduced in size to accommodate clumps of trees in retention, we must consider other options and we may not have enough new equipment, if we convert to a variable retention system. A surprising finding from our models is that harvesting options such as helicopter and skyline logging have become more economical; regulations have reduced sizes of clear-cut sites to the point where the cost of road building is only slightly below the cost of other logging practices on many types of terrain; sometimes, it is more expensive. In variable retention, with cut sites much smaller, the cost differential is diminished further. Logging for value instead of volume, as noted above, may offset some of the costs.

In sum, we estimate that variable retention logging will add about 5% to the cost of cutting trees. While lumber prices fluctuate with the market, we based our estimates on an average of CAN$400/m^3 in sales, and profits at 5–7% of sales at CAN$24/m^3. Cost and other risk factors include: availability and cost of equipment; start-up costs; time to retrain loggers; and unions reluctant to agree to variable retention. Benefits and potential cost recovery would come from technological improvements, higher prices, productivity improvements and better access to trees due to fewer conflicts with environmental groups.

Box 12.6 Report of the profit subcommittee to the forest project team (continued over)

> **Recommendation.** Cost increases run counter to current efforts to cut costs and improve efficiency. We cannot expect a green premium for changing to variable retention, which means any cost increases must be recovered through other means. There is the possibility to make variable retention work economically, but it is a very high-risk strategy.

Box 12.6 (from previous page)

The Forest Project team disagrees

After the reports were presented, lively discussion ensued and generated some additional considerations:

- Harvest levels in the past few years had declined by 20% because of environmental controversies, and were currently holding at 5.7 million m³. This number could be expected to go down further if no environmental solution was found. If the environmental controversy died down, it would be conceivable that the company could access all 6.2 million m³ of its potential annual allowable cut, though of course that amount would have to be reduced by whatever amount was committed to old-growth preservation or variable retention

- Some of the wood that was usually removed from the clear-cut site was waste wood (e.g. deadfalls); it currently did not count in cut calculations, but needed to be removed from the site because of government guidelines. This wood was actually important habitat for species, and thus should be retained on-site. It was estimated that considering this waste wood could make up approximately 4% of that retained. Furthermore, about 5% of the total area (1.1 million hectares) logged by MB was already preserved owing to forest practice code regulations protecting areas alongside stream beds (important salmon spawning grounds) and on hillsides (because of erosion issues)

- Current regulations required that not only the best trees, but also the poor trees were taken out of the forest. The poorer trees might be those that had decay or timber 'defects' such as large knots or branch structures that reduced their timber value. Trees that were too small to yield timber of a size demanded by the market also had to be cut. All of these trees were uneconomic to cut. For preserving biodiversity and habitat, however, they were often as well suited as any other tree. Leaving such trees was a win–win for habitat and economics. Special habitats such as ravines, wetlands and rock outcrops typically had higher plant diversity, yet lower timber values. Leaving these sites as part of permanent reserves would help maintain biodiversity. Uneconomic trees accounted for an average of 5% of the total harvest. Not cutting those trees would save the company a significant amount in handling costs

Cafferata ponders the decision

Cafferata thought again about his options and what he and the Forest Project team were going to do next. A complete halt to cutting old-growth forest would virtually be the end of MB in BC, since there would not be enough volume left to sustain operations. However, if the company failed to make any move on old growth, environmental groups would be unlikely to stop the customer campaigns and to give MB some peace. On the other hand, moving to a variable retention system would answer the environmentalists' demands to end clear-cutting, but it would involve significant investment in new harvesting equipment and employee training. It might also be unpopular with loggers who had spent their lives clear-cutting, and it was sure to be unpopular with the other forest companies. Some accommodation by the government would also be necessary to ensure that the changes did not result in charges of violating the Forest Practices Code.

One thing was sure: MB had to do something to get the environmental monkey off its back, and a strong move might give the company a competitive advantage with customers, at least in the short term. It might be costly, however. A weak move could intensify the pressure on MB, if environmentalists cried 'too little, too late'. In the long term, the pressure was sure to return no matter what decision was made, as there were many environmentalists who wanted no logging of old-growth timber in BC at all.

Thinking back to the 90-day review process, Cafferata remembered Tom Stephens's admonition to business unit managers to 'sell it, milk it or grow it'. Milking the forest assets would involve cutting and selling as much as possible, until social or government forces moved to stop them, then exiting the business. The company's reputation would be likely to suffer from a milking strategy but, if the company exited the business, it might not matter. On the other hand, selling the assets immediately could be a viable option. Although the assets would not be likely to fetch a high price given the 'BC discount', the company's market capitalisation currently valued them at zero: the stock price reflected only the value of the non-BC assets. Selling would bring in some cash that could be used to grow the business in other areas, and would be likely to increase the company's stock price and get the investment community off MB's back.

If the Forest Project team did recommend keeping the assets, they needed to present a unified front to the board of directors when they made recommendations—and they needed to present a decision that was acceptable to the board and the institutional investors. At this point, that seemed only a distant possibility. Cafferata could take the team's input under advisement and then make the decision himself as to what to recommend to the board. Or he could use his power as the team leader to push people in one direction. However, the people on the team were just the starting point. If he couldn't get agreement among these people, what hope did he have of getting buy-in from all the other parties involved: the employees and unions, the government, the environmentalists, customers and aboriginal groups?

Part IV
Role-play simulations

13
Sustainable games people play
TEACHING SUSTAINABILITY SKILLS WITH THE AID OF THE ROLE-PLAY 'NORDWESTPOWER'*

Anke Truscheit and Christoph Otte
University of Oldenburg, Germany

Development background

The role-play 'NordWestPower' is a behaviour-oriented role-play developed by teachers from the Department of Economics and Law at the University of Oldenburg.

There were two starting points for the development of this role-play:

- The main goal was to train students to be sustainability managers. For this, it is essential not only to impart specialised knowledge, but also to train methodical, communicative and social skills (Freimann and Schwaderlapp 1994; Greimel 1999). Our graduates should be able to communicate and work together across interdisciplinary borders

- Conventional teaching styles such as lectures and seminars alone are not appropriate for teaching these competences (GiMA 2000). Research over the last few years has shown that role-play exercises are a particularly successful tool for teaching the so-called 'soft' skills (Rebmann 1998)

During the training of sustainability managers, the teaching of soft skills gains special importance. This is because the complexity of both sustainability issues and vocational

* We thank Thorsten Möhlmann and Uwe Schneidewind for their dedicated co-operation during the development and realisation of 'NordWestPower'.

reality cannot be handled successfully with technical knowledge alone. As a result, one can divide the additional sustainability skills into three groups (see Table 13.1).

Methodical competences	Social competences	Individual competences
● Decision-making techniques	● Teamwork	● Creativity
● Project management	● Integration	● Flexibility
● Presentation	● Conflict resolution	● Networked intellect
● Learning techniques	● Motivation	● Leadership qualities
	● Communication	● Self-reflection

TABLE 13.1 **Sustainability skills**

The best approach to training managers in these skills is to use experiential study pedagogies. These include role-plays and active learning methods, which enable experimental and playful learning (learning by doing, action learning). The conditions of work for the participants in the role-play are comparable with the real world, characterised by teamwork, decisions under time pressure and imperfect information. In order to reach this, either an enterprise as a whole or separate functional areas of an enterprise are simulated. The participants of the role-play act as employees of the enterprise. In their roles, they are confronted with hypothetical problems which must be solved as part of a team. The participants are thus forced to make decisions. They experience the results of their decisions without having to undergo the corresponding real-life consequences.

The structure of the role-play 'NordWestPower'

NordWestPower is a behaviour-oriented role-play. This means that participants are not subject to some formal, computer-based algorithm. Instead, they have a broad scope of action at their disposal with which they can gain methodical, social and individual competences. Only the essential features of the experiential scenario are laid down by the direction of the role-play. The basic structure of the role-play is shown in Figure 13.1. Within the planning game, a certain scenario is given: that is, framework defaults and framing data for NordWestPower. The role-play execution is observed and controlled by the teachers. If it deviates extensively from the default setting, the instructor may choose to intervene and to provide support to the participants. This is discussed in further detail below.

Participants are divided into three groups which form the enterprise's 'departments'. These departments are: marketing, controlling and legal. The director can be played by either the instructor or one of the students.

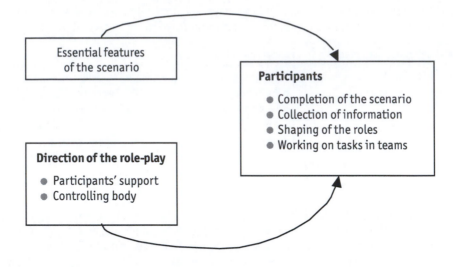

FIGURE 13.1 Basic structure of behaviour-oriented role-plays

The course of role-play in 'NordWestPower'

The whole play takes one semester, and 15–20 students can participate. It is advisable to use an internet platform (e.g. Yahoo Groups) during the play. This allows all participants and teachers to communicate with each other, and documents, presentations and interesting links can be posted.

The role-play starts from the following situation: The company 'NordWestPower' is faced with the challenge of developing a new long-term strategy. It's obvious that the use of renewable energies will play an important part in this strategy, but management still requires information and advice about this strategic direction. For this, they decide to set up an internal consulting team consisting of the participating students. Every student plays the role of an employee who comes from a special department of the company (marketing, controlling, legal). Figure 13.2 shows the various phases of the role-play.

FIGURE 13.2 Course of the role-play 'NordWestPower'

The application phase

In the first phase, students must apply for participation in the role-play. Participants 'apply' to the internal consulting team and include a CV, description of their goals, suitability to the project and so on. The application should also indicate the department in which the participant would like to work. Students are selected based on their applications. The application phase achieves two goals: (a) creating a realistic start to the planning scenario; and (b) teaching participants appropriate techniques for applying for sustainability jobs. These experiences are useful to students on entering the working world of sustainable business.

The seminar phase

In the seminar phase the students get some basic information about their company, the energy industry, current energy policy and law in Germany and the European Union, climate policy and resource political background. Of course, the seminar topics would be different if the exercise took place in a different geopolitical context.

Currently, the energy industry stands at the centre of exciting political and social discussions. The creation of a sustainable energy supply is one of the key pillars needed for a sustainable economy (Hofer and Scheelhaase 1998). Within the range of the energy industry, the connections between economic, ecological and social issues can be studied intensely. Discussions on phasing out nuclear energy, taxation of energy consumption in the context of ecological tax reforms, and the promotion of renewable energies against the background of climate protection aims are the current highlights (Bundesministerium für Wirtschaft und Technologie 2002; Enquete-Kommission 'Nachhaltige Energieversorgung' des 14. Deutschen Bundestages 2002).

During the seminar phase we recommend bringing in guest lecturers. In order to prepare students for their roles, it is very useful to arrange meetings with experts from local energy companies who report on their daily business and explain the strategies of their companies. During this phase, it is possible to organise excursions and visit different power plants (nuclear, coal, renewable). Our past experience has shown that these visits help students understand the technical background of electricity production and distribution.

As part of the seminar phase each participant writes a working paper of approximately ten pages. These papers are made accessible to all consulting team members via the internet group platform and serve as a common information basis for the strategy development process. As part of this strategy formulation process the participants decide which subjects need more detailed preparation and analysis. We suggest that the role-play director does not interfere as long as no key central topics are forgotten. In the following, we will discuss some of the topics discussed in these papers.

The following questions are posed: How can NordWestPower further bind their customers to the enterprise? With which strategy can potential customers be acquired? The marketing department must develop a marketing concept. Using the example of the task 'comparison of the profitability of wind energy power plants' the controlling department points out the fact that regenerative energy represents not only ecological advantages but also an economically safe investment for future NordWestPower enterprises.

The strategy weekend

After the seminar phase the strategy weekend takes place. This weekend is the role-play's highlight for most students. The normal classes are placed on hold, and the weekend is spent outside the university in an environmental education facility or something comparable. It is often desirable to begin the weekend with a common outdoor group event, such as a canoe trip. This combines the positive effects of group dynamics with a direct experience of nature and the environment. Afterwards, the representatives of the different departments work in separate parallel meetings on their department's respective position on the elements of the strategy. During these meetings, each group is observed by a role-play director who stays in the background, observes the discussion behaviour of the participants, but does not interfere in the group processes. Every small group has to prepare a short presentation for the common strategy meeting of the whole consulting team. Therefore, it is helpful to equip the groups with some technical support such as a laptop with presentation software.

The common strategy meeting comprises the last part of the weekend. Every department presents its contributions to the future strategy of the company. The whole group has to co-ordinate these parts in order to develop a self-contained draft strategy which is supported by each participant. This element of the role-play aims especially at dealing with conflict situations. The participants are forced to search for compromises between conflicting positions: for example, they have to solve the conflict between the positive marketing effects of green electricity and its higher production costs. The meeting is observed by the role-play directors, as well. They register the discussion behaviour of every participant and the way the group resolves their conflicts: Is the group dominated by some members? Do they reach an agreement by democratic voting with the minority accepting the results? Or do they search for a compromise that everybody can accept?

After the strategy weekend, the participants prepare the final presentation to the NordWestPower board of directors. The board is formed by the role-play directors, representatives of local energy companies, and faculty staff members who have some knowledge of energy and sustainable development. The competence of the board and the examination-like situation create a real challenge for the students. They must perform a convincing public presentation and defend the results of their work of the past months against hard criticism. In order to guarantee a thorough assessment of the students' performance, the whole presentation is recorded on video.

The evaluation phase

The role-play ends with the evaluation phase, which has two components. First, the participating students assess the concept and course of the role-play and the work of the role-play directors. This is done with the help of an anonymous questionnaire and as part of a final discussion. Next, in a discussion with a role-play director, every student receives detailed personal feedback assessing the quality of the contents he or she delivered during the simulation and evaluating the manner in which he or she acted during the discussions and presentations. This feedback is based on the application, the working paper, the director's notes during the strategy weekend and the video recording of the final presentation. The sustainability skills mentioned are the focal point of this feedback. In case a grade is necessary, this is awarded on the same basis.

Some concluding remarks

Role-plays are an innovative didactic teaching instrument for different reasons. From the view of the students, the relevance of instructional contents becomes clearer in a role-play. Our experiences showed that the acceptance of the role-play is very high with both students and teachers. This is because the character of the role-play offers a high degree of motivation for both groups even if it is much more labour-intensive for students and instructors. Our experiences from previous years bear this out as they have been very positive.

In Oldenburg, the role-play is integrated into the basic Ecological Economics course of study. Second-year students are eligible to participate. Because of the open, behaviour-oriented design of the role-play, integrating it into other phases of economic courses with an environmental or sustainability focus is easily accomplished.

We believe that it is crucial to teach sustainability skills. Sustainability managers in companies, advocacy groups, NGOs (non-governmental organisations) and government must handle many varied issues and challenges. For their work, they need excellent competences in the area of teamwork, communication and conflict resolution. The realistic, experiential role-play, NordWestPower, helps to teach these skills and competences and does so in a stimulating, challenging environment.

Note

If you are interested in the role-play and want to use NordWestPower at your educational institution, or if you need help in developing your own role-play, please don't hesitate to contact the authors.

Bibliography

Bundesministerium für Wirtschaft und Technologie (2002) *Sustainable Energy Policy to Meet the Needs of the Future* (Berlin: Bundesministerium für Wirtschaft und Technologie).

Enquete-Kommission 'Nachhaltige Energieversorgung' des 14. Deutschen Bundestages (2002) *Nachhaltige Energieversorgung unter den Bedingungen der Globalisierung und der Liberalisierung* (Abschlußbericht; Berlin).

Freimann, J., and R. Schwaderlapp (1994) 'Fast wie im richtigen Leben: Erfahrungen mit dem Verhaltensplanspiel "Pappenheim" ', *Zeitschrift für Betriebswirtschaft* 2.94: 53-64.

GiMA (2000) 'Gesellschaft für integrierte Management-Ausbildung mbH', homepage, www.gimaconsult.com, 30 June 2006.

Greimel, B. (1999) 'Das didaktische Potential von Unternehmenssimulationen: Empirische Befunde und theoretische Reflexion', *Wirtschaftswissenschaftliches Studium* 3: 156-60.

Hofer, P., and J. Scheelhaase (1998) *Nachhaltige Entwicklung im Energiesektor? Erste Deutsche Branchenanalyse zum Leitbild von Rio (First German Sector analysis after the Rio Example)* (Heidelberg, Germany).

Lachnit, L. (1996) *Absolventenbefragungen als Instrument zur Beurteilung der universitären Ausbildungsleistung. Bericht zur Absolventenstudie des Fachbereichs Wirtschafts- und Rechtswissenschaften der Carl von Ossietzky Universität Oldenburg* (Oldenburg, Germany).

METEOR (2000) 'METEOR-Zentrum für studienbegleitende Schlüsselqualifizierung an der Fach-hochschule Mannheim', homepage, www.keysoftware.de/Meteor-BV, 30 June 2006.

Rebmann, K. (1998) 'Der Planspieleinsatz aus der Sicht von Lehrern und Lehrerinnen sowie Ausbildern und Ausbilderinnen', *Zeitschrift für Berufs- und Wirtschaftspädagogik* 94.4.

14
Using experiential simulation to teach sustainability

Susan Svoboda

Realia Group, USA

John Whalen

Sustainable Value Partners, Inc., USA

Creating sustainable business solutions requires seeing a business as part of a much larger system involving a wide variety of stakeholders—more like the complex interactions in a natural ecosystem. Leadership in this complex and dynamic context requires a perspective that incorporates social and environmental dimensions as well as economics, and requires a flexible, adaptable and inclusive approach. Results-oriented MBAs and executives often prefer systems that emphasise efficiency and order rather than adaptability and openness. They may struggle with how to integrate dynamic natural systems and complex social processes with control-oriented management systems. The perspective and skills required for this integration can only be learned through experience. This makes simulations an ideal medium for teaching sustainability.

This chapter is a practical guide for using experiential simulations to teach sustainability in a business context. The chapter draws on the authors' experience in developing and using a simulation exercise called Transformation with hundreds of groups in business schools, corporations, government organisations and non-profit organisations over the past six years. The chapter describes the benefits of using simulation in building understanding of sustainability, describes the Transformation simulation, discusses typical lessons learned, and provides a checklist of the characteristics of an effective sustainability simulation.

Experiential learning and transformational change

Experiential learning is one of the most effective ways to promote positive change in individuals and organisations. The experiential learning model consists of cycles comprising four basic steps: Act, Reflect, Reframe and Apply.

- **Act**. Experiential learning is based on actions and their observable results as the basis of learning

- **Reflect**. Experiential learning provides an opportunity for participants to get feedback on their actions and explore the results, as well as discover mental models

- **Reframe**. When participants gain understanding of the impact of their actions they can change the mental frames that prevent them from achieving the results they want

- **Apply**. Experiential learning makes learning transfer explicit, building clear linkages between the insights gained in the 'artificial' learning process and the real-world challenges facing the participants back on the job

Successful experiential learning processes usually run through this cycle two or more times to deepen integration of the learning. This enables participants to experience the results of acting within their current frame of reference and mental models, gives them an opportunity to reflect on and analyse these results, and experiment with shifting their behaviour and seeing how the results change based on the shift. They can then reflect on how to best apply their learning to the real world in the form of new ways of thinking and acting in their jobs.

One of the sources of the power of experiential learning is that it engages the whole person, involving the participant's mental, emotional and somatic intelligence. You might say experiential learning treats the person as a complex living system! In this sense the medium is the message: experiential learning may be the most ecological kind of learning experience other than real-life experience itself.

Business simulation as a mechanism for teaching sustainability

Business simulation is a type of experiential learning that is well suited to teaching sustainability for several reasons:

Complex systems interactions

One goal of sustainability education is to raise people's awareness of and understanding of complex systems. If participants don't understand the system in which they are working, they may take actions that produce unintended consequences. For example, without an understanding of the product life-cycle, a team may make a product design change to eliminate a hazardous substance in the manufacturing process that creates more pollution/risk when the consumer uses the product. Simulation places participants in the middle of a complex system and enables them to experience the impact of their actions on that system in a low-risk environment.

Cross-group collaboration

A simulation bridges differences in expertise, professional languages and cultures by drawing on each individual's accumulated knowledge and skills. At the end of the cycle, participants have an opportunity to reflect on and discuss their results, learning with colleagues from other backgrounds. To build truly sustainable solutions, collaborative learning is an essential aspect to working with stakeholders with multiple and conflicting interests.

Integration of economic, environmental and social dimensions

By combining traditional indicators such as income statements and balance sheets together with sustainability benchmarks, such as pollution or impact/product, a simulation can enable students to explore how to increase profits by optimising their use of natural and human capital. This integrated approach is much more powerful than attempting to teach business people about sustainability as an issue separate from concerns for profitability, market share and revenue. A simulation is an ideal vehicle to demonstrate the interconnections between these dimensions in a relatively short amount of time.

Consequence-free practice

The simulation's hypothetical context enables participants to disengage from their existing paradigms and open themselves to broader perspectives. It provides a rare opportunity to experiment and try new behaviours in a relatively consequence-free environment with immediate feedback on what is working and what is not. If we want managers to make the best decisions when it really 'counts', we have to give them a chance to practise.

The Transformation exercise

Transformation is a reality-based, team-building simulation that helps participants understand how to translate the concepts of sustainability into tangible action. The Transformation simulation uses Stuart Hart's sustainability portfolio (described in Hart 1999) as the framework for categorising sustainability issues and the strategies by which teams in the simulation compete. Teams represent companies that produce products, and individual team members take on roles within their companies. Each team makes a product and runs their company in the context of lifelike conditions such as time pressures, budgetary constraints, unpredictable stakeholder interventions, changing market conditions and limited information. At the end of the decision cycle, participants sell their product in a dynamic market that allows companies to take market share from one another, and record their decisions, which are 'scored' electronically.

The Transformation scenario addresses sustainability in the broadest meaning of the term—the companies must compete in an environment where success depends on their ability to select a market, create a product that is attractive to that market, design and deliver that product profitably, and deal with social and environmental issues that may increase their costs or hinder their ability to compete.

The roles

Participants are assigned to functional roles in their company, including marketing, product design, manufacturing, finance and environmental management. Each role has an instruction manual that explains the goals of the exercise, team and individual objectives and scenarios. As in the real world, each member of the team has some shared and some unique information and goals.

The process

Participants come together as a team, select a market, design and build a product, and create a marketing campaign for that product, while competing with other companies. They must also complete market and profitability analyses. The 'product' is made using K'nex toy construction components, balancing the economic, environmental and social impact of the components. The product is evaluated on how it meets several performance criteria and the company is evaluated on leadership and management issues such as being proactive, transparent and innovative. Reflecting the real world, the management teams often find they are working with incomplete, changing or conflicting information. The facilitators play the part of different stakeholders, influencing the proceedings proactively and in response to company actions. These stakeholder interventions allow the exercise to be customised to address particular issues of importance to the participants. It also enables us to reshape the exercise at the spur of the moment, compelling participants to react to changing circumstances.

The results

For the purpose of evaluation, participant companies form an industry comprising five to nine company teams, the number that experience has proved to be optimal for learning. Proprietary software tracks company decisions and generates income statements, pollution reports and balance sheets while facilitator evaluations offer qualitative feedback on team performance, stakeholder interactions and industry dynamics. Each industry comprises 15–40 participants, and groups larger than 40 participants are divided into multiple industries.

Multiple levels of learning

Participants in a simulation operate at several 'nested' levels simultaneously, and therefore have rich opportunities to learn on all these levels.

Personal

At the personal level, the simulation teaches participants about their assumptions about how the world works, and about how they work in the world. For example, some participants are surprised to find that they did not seek out new sources of information, did not listen to their customers, or assumed that a 'green' product would have to cost more than a traditional product. They learn about their own behaviour and the mental models it reflects.

Interpersonal

At an interpersonal level, the simulation teaches participants about how they relate with other individuals on the team, and how those patterns of relating support or hinder success. For example, they may withhold viewpoints until it is too late in the process, and find their perspective would have prevented a major problem if they had been more assertive. Or they may find themselves taking more risks than they usually do in 'real life' to reach out and build trust, and see that this creates positive results.

Role

The simulation can enable participants to see a situation from a new functional perspective, as well as illustrating how an issue cuts across all functions. For example, when participants with an environmental background take, say, the finance role, they quickly learn why the finance department is concerned with its issues, and how to relate in a way that is more meaningful and relevant with these folks in the future.

Team/organisation

Today, most work is done in teams, so learning should be done that way, too. The simulation shows how to work more effectively as a team by aligning individual practices through sustainability strategy. For example, common lessons include the tendency to assume buy-in or support of other parts of the organisation without doing anything to gain it. Also, teams frequently realise that, unless their sustainability practices align with their business model, the commitment to sustainability will be sacrificed to other business priorities.

Industry

Simulation gives participants a chance to experience how sustainability strategies can produce marketplace advantage, if they are implemented in a way that creates more business value than their competitor. For example, over the course of a workshop, practices that early in the day appear innovative—such as pollution prevention—lose their competitive edge as every team in the industry adopts those same practices. As the bar rises, teams must look for new ways of creating sustaining value, including stakeholder dialogue, clean technologies and rematerialising.

Lessons learned in the Transformation simulation

Through the simulation, participants learn to see sustainability as an integral part of the business model, and experience how this new frame can lead to business success. This is a major shift from viewing sustainability issues as external to the enterprise's sphere of interest and influence. As part of this shift, participants learn the following:

The importance of listening to customers

About a third of the teams fail to talk to customers during the concept and design process. This results in activities that tell customers what they 'should want' rather than listening to them and delivering what the customer really wants. For example, some teams decide that, since this is a 'green' simulation, customers will buy a product that is environmentally 'correct' even though it sacrifices performance. On the other hand, those teams that create a real dialogue with customers position environmental characteristics in terms of value perceived by customers: for example, they will talk about a durable product that offers the customer reliability and convenience while creating less waste.

The benefits of stakeholder engagement

Most teams begin the exercise thinking of stakeholders as a nuisance that must be managed to ensure that they do not interfere with the efficient operation of the business. A key learning for most participants in Transformation is that stakeholders, when engaged, have valuable knowledge and perspectives that can help them to achieve their goals more quickly and with higher-quality results. Some teams go through a superficial communication process with stakeholders, asking them for input at the start of the exercise but not engaging them in continuous dialogue or incorporating their ideas into the team's strategy. These teams almost always lose to competitors who integrate stakeholders in a more comprehensive way into their process.

The importance of vision and values

All Transformation teams start by writing a brief vision and values statement that is posted for all to see. When teams run into trouble, such as running out of time or money, very few return to their mission and values statements to help guide them through the process. In some cases, their actions—such as accepting substandard pollution or human rights impacts in their facilities—are clearly not in keeping with their mission and values. When these teams later reflect on and discuss their performance, they are surprised at how quickly they discarded their vision and values in the effort to get a profitable product to market on time.

The value of relationships

Many innovative solutions to achieve sustainability require significant changes in thinking and/or co-operation of multiple parties. Many teams have a tendency to assume they have buy-in and support from other parts of the organisation without doing anything to gain it. They come up with innovative solutions, but forget to build the support they need to implement them. This is especially evident when teams are moving into new and emerging markets. On reflection they see that their inability to build effective relationships was the downfall of their strategy.

The importance of adaptability and flexibility

Many teams focus so intensely on achieving the goals they have set that they refuse to revisit decisions based on new information. When they run into a roadblock or barrier they simply push harder, rather than taking a flexible and adaptive approach. For example, teams will observe one of their competitor's breakthrough designs and simply ignore it and continue their own struggling project, when they could easily learn from their competitor's success.

Challenges and limitations

One challenge of using a complex simulation such as Transformation is the time commitment. Transformation requires a minimum of a half-day, and ideally a whole day, plus pre-reading by the participants. A whole-day session enables the participants to go through two complete product development cycles, allowing them to create a new strategy and product based on what they have learned. Two- or three-day sessions allow participants to combine readings and case studies with three cycles of simulation for more in-depth learning. Getting managers in a corporate setting to allocate this much time for such an exercise can be difficult.

Another challenge is that some participants may have a bias against experiential learning, believing that they can't learn anything important by 'playing games'. This is typically a problem in getting people to attend, but not during the simulation itself. The process of experiential learning in a well-designed simulation is so engaging that most sceptics are quickly swept up in solving the business problem.

A further challenge is the multiple levels of activity that must be facilitated. Keeping an eye on the work process, the interpersonal dynamics, and co-ordinating the logistical details of the exercise requires a skilled facilitator. The experiential nature of the exercise, however, creates a largely self-regulating system; real-time feedback from team members, competitive pressure from other teams, and the pressure of the clock help keep participants focused and working effectively.

Although simulation is in many ways much more 'real-world' than other forms of learning, it is still not the real world. As in all kinds of learning, transfer of skills and knowledge to the real work environment is the biggest challenge. Simulation has a great advantage in that participants are not simply gaining knowledge, but are actually putting knowledge into action—demonstrating behaviour and getting real-time feed-back on that behaviour. Ensuring transfer of learning from these experiments back to the work environment is best facilitated by having participants reflect on what they learned and what they will do differently back on the job based on that learning, and developing an action plan they can implement based on their learning. A follow-up session some months after the simulation can provide an additional learning cycle and reinforce the new behaviours.

Implications for sustainability educators

Simulation is a powerful way to disconfirm old beliefs and mental models and open participants to new ways of thinking about sustainability. Thus it can be a great way to begin a sustainability course, or to introduce a sustainability planning session, or to build a foundation for more effective teamwork and innovation within an organisation. The whole-person learning that occurs through a good simulation can create an openness to other forms of teaching and can lay the groundwork for a different level of dialogue and creative collaboration.

Developing simulations is a complex, time-consuming and expensive process. For this reason, many people prefer to use an existing simulation, rather than creating a new one. The following checklist can serve as a useful guide in evaluating effective sustainable business simulations. We suggest you look for ones that will:

- Provide a rich and manageable model of the real-world complexity of business, with the business system realistically integrated with social and natural systems that interrelate and affect each other

- Offer an appropriate level of challenge to keep business people interested and engaged

- Be realistic enough to model the complexity of the real world but not too complex or confusing to enable learning in a short time-frame

- Be flexible enough so participants can bring their own knowledge of the latest techniques

- Encourage innovation in technology and behavioural change

- Illustrate both the technical aspects of sustainability as well as the communication objectives

- Be based on a tested framework for thinking about sustainability

- Present an opportunity for participants to take on one or more roles or perspectives that are different from their usual ones

- Be anchored in things that are important to business people (profitability, market share, innovation, return on investment) and demonstrate how sustainable choices are relevant to these measures

- Prevent participants from failing, yet provide enough challenge to promote learning and use the exercise as a safe environment for growth

- Provide learning on multiple levels (personal, team, organisation, industry)

- Supply clear and easily understood procedures that ensure a logical progression of activities, including steps or play, cycle critique, rules and policies

- Provide two or more loops through the experiential learning cycle

- Present feedback that focuses on the most important aspects of business sustainability

- Use indicators and an accounting system that reflects tangibly and realistically how a situation or decision has an effect

- Afford an opportunity for clear mapping of the lessons learned back onto the real-world environment through critique and debrief discussion

Reference

Hart, S. (1999) 'Beyond Greening: Strategies for a Sustainable World', *Harvard Business Review*, January 1999.

15
Teaching process sustainability
A ROLE-PLAYING CASE FOCUSED ON FINDING NEW SOLUTIONS TO A WASTE-WATER MANAGEMENT PROBLEM

David Annandale and Angus Morrison-Saunders

Murdoch University, Australia

When people think about sustainability, they often focus on 'content' issues. It is natural for people to want the outcomes of decisions to be sustainable: for example, for organisations to reduce their production of wastes, or for individuals to consume energy that is produced from renewable sources. What is sometimes forgotten, however, is that the *process* by which we move towards sustainable outcomes also needs to be sustainable.

This chapter examines the link between sustainable outcomes and process sustainability, by way of a case study. The 'outcome focus' of the case is the attempt to determine a new, and more sustainable, waste-water treatment option for a regional town in a particularly beautiful part of southern Western Australia. Its inhabitants have a strong interest in environmental preservation. In the early 1990s the Government Water Authority announced its intention to construct a new secondary waste-water treatment plant that would emit treated sewage from an outfall pipe on a cliff overlooking the Southern Ocean.

This interactive, role-playing case study investigates the problem by assigning stakeholder roles to students. The case allows students to reassess the way government bureaucrats made their original decision, and has them search for a better solution. It asks students to think about whether it is possible to find a sustainable solution to a problem, without addressing the sustainability of the decision-making process.

The chapter begins with an introduction to the content of the problem, and then outlines the development of conflict surrounding possible solutions. It moves on to introduce the concept of 'process sustainability', and then outlines the interest groups involved in the controversy. Finally, it presents some information on the timing of the development proposal, and finishes with some conclusions and a collection of background material that further illuminates the problem.

In large part, the chapter has been designed to allow for interactive role-playing, where students can participate in a public meeting and work towards resolving a conflict. Ideas for instructors are included as an attachment at the end of the chapter.

Background

The town of Albany, situated 410 km south of the capital city of Perth, is an economic focus for Western Australia's Great Southern region. Historically, the region's economy has been dominated by primary production, largely agriculture and fisheries, and in recent times it has become increasingly dependent on tourism.

The natural environment of the Albany district is a major attraction, and a reason for inward migration. The pristine, often rugged coastline is also prized for its tourism value and the region is marketed as the Rainbow Coast. Not surprisingly, Albany residents have high expectations for the quality of their surrounding environment and are thus especially sensitive to any developments that may affect it.

During the 1980s the Environmental Protection Authority (EPA) undertook a major study into the impacts of industrial, agricultural and residential developments on Albany's harbours (EPA 1990). That study showed that waste products from a variety of activities were elevating nutrient levels within the harbours to values that could not be assimilated by the system without leading to adverse effects. Consequently, the biotic environment of the harbours was being severely degraded. Part of the pollutant load entering the harbours was identified as the Water Authority of Western Australia's (Water Authority's) treated domestic sewage from Albany.

The EPA's report carried a number of recommendations, one of which was that the Water Authority should cease outflow from its 36-year-old King Point Treatment Plant (close to Middleton Beach) by 1994 (see Fig. 15.1). At the same time, around the end of the 1980s, the Water Authority had also realised that continued population growth was placing undue strain on ageing infrastructure and that the time had come for the upgrading of existing waste-water treatment facilities.

Sewerage development at Albany

A reticulated sewage collection, treatment and disposal system was first constructed in Albany in the 1960s to alleviate major health hazards. Since then, the system has undergone a series of modifications and expansions to meet the needs of a steadily growing number of users. During the period in which this case study developed—the early

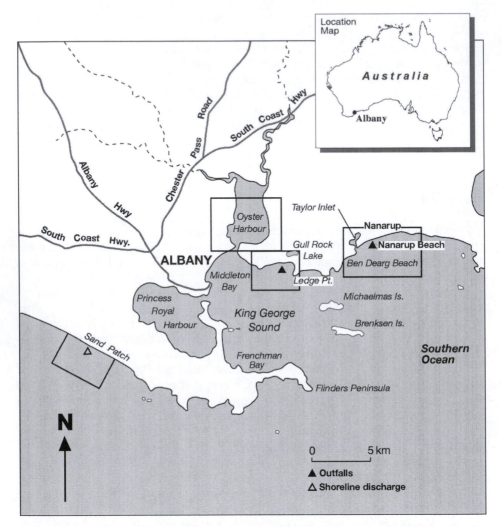

FIGURE 15.1 Proposed location of options for marine disposal of waste-water for Albany

1990s—it consisted of four waste-water treatment plants, each with its own disposal system.

The King Point waste-water treatment plant (No. 1) handled the majority (approximately 70%) of Albany's sewage: in mid-1988, this represented 2,900 domestic connections (servicing about 8,700 people). Sewage here underwent primary treatment and treated waste-water was discharged to nearby King George Sound.[1]

The Timewell Road waste-water treatment plant (No. 2) treated the bulk of the remaining effluent: in July 1988, this represented 1,150 domestic connections (servic-

---

1   Primary treatment involves screening and sedimentation to remove solid materials from the waste-water.

ing about 3,450 people), plus the discharge from Masters Dairy (equivalent to that of about 830 people). Secondary treated waste-water from this plant was discharged through a small wetland watercourse into Five Mile Creek, eventually leading to the Southern Ocean through the Torbay Inlet.[2]

Remaining waste-water treatment plants (No. 3 and No. 4) handled small amounts of effluent—sewage from 400 and 220 people respectively, in 1988. After treatment with activated sludge, reclaimed waste-water from these plants was disposed of on-site into sandy soils.[3]

A major internal review of the Albany sewerage system found that the capacity of all plants to expand in response to increasing flows was extremely limited. The review developed and compared seven treatment and disposal options to cater for the increase in demand into the next century. These options covered a variety of waste-water treatments at single or multiple plants, and a range of disposal options including discharge to ocean, inland watercourses and land. The review's final recommendation was that effluent be treated to secondary level by aerated ponds both at No. 2 waste-water treatment plant (Timewell Rd) and at a new plant to be sited near Cuthbert (to be No. 5 waste-water treatment plant), and then be discharged to the Southern Ocean at a site 700 m west of Sand Patch (see Fig. 15.1). Plants numbers 1, 3 and 4 would be closed down.

Development of conflict

Early in 1990 the Water Authority released the findings of its internal review to the public, in the form of a 'glossy' brochure promoting the cliff-edge discharge option at Sand Patch. The publication of this decision raised considerable opposition from a range of interest groups, including surfers, salmon fishermen, abalone divers, environment action groups and the general public. Such was the concern that the Water Authority decided to undertake a programme of community consultation to assess the overall community reaction to the proposal. It established a community-based committee under the chairmanship of retired Murdoch University Environmental Science Professor Des O'Connor.

The O'Connor Committee was given access to the Water Authority's internal review. The seven options investigated by the Water Authority included:

- Secondary treatment and dispersal to Sand Patch, Nanarup Beach or Ledge Point (see Fig. 15.1)

---

2 Secondary treatment involves a biological treatment process, following primary treatment, in which organic wastes are consumed by bacteria under controlled conditions.
3 Activated sludge derives from the bacterial consumption of organic waste during the secondary treatment process. Aerated bacteria from the sludge converts the organic component of waste-water into bacterial mass and stabilised compounds such as nitrates, sulphates and carbon dioxide.

- Tertiary treatment[4] and dispersal to Sand Patch

- Secondary treatment and dispersal to land within a reasonable radius from Albany

Following a number of public meetings and the receipt of submissions, the O'Connor Committee—with public support—suggested that the Water Authority pursue a 'zero' discharge option where fully treated waste-water would be recycled and used again for domestic purposes. This suggestion was essentially an extension of the tertiary treatment option, without the need for a disposal outfall. The Water Authority refused to accept the 'zero' discharge option, believing that it would be too expensive and could not be justified given the relatively small amount of waste-water requiring disposal, and the turbulence of local ocean environments.

Towards the end of 1990, opposition became even stronger, as local people's perceptions of Water Authority intransigence hardened. It became clear at this point that if the Water Authority continued to pursue secondary treatment and pipeline disposal at Sand Patch then local opponents would fight the decision, possibly in the courts.

What is 'process sustainability'?

While the definition of 'sustainability' is still problematic, it is fair to say that there is a growing understanding that sustainable outcomes need to somehow balance the 'triple bottom line' of economic, social and environmental goals. The Water Authority's choice of secondary treatment and pipeline disposal at Sand Patch could well be viewed as a sustainable outcome from the point of view of the Authority. It clearly leads to a better environmental outcome than what went before, and it appears to be economically attractive.

As the background material presented above indicates, however, it would be difficult to consider the Water Authority's decision as being sustainable from a process perspective. Two tactics—the Authority's initial 'decide–announce–defend' approach, and the later more open O'Connor committee—both resulted in a dissatisfied community. This suggests that a good (i.e. sustainable) outcome might be reached only if a sustainable *process* is used to guide decision-making. If we want to design sustainable decision-making processes, it would help if we had an understanding of how they have been defined in the literature.

A recent approach to establishing a Sustainability Strategy for Western Australia presented seven 'foundation sustainability principles' and four 'process principles'. The process principles were: integration of the triple bottom line; accountability, transparency and engagement; precaution; and hope, vision, symbolic and iterative change (Government of Western Australia 2002).

---

4   Tertiary treatment is the further treatment of effluent from the secondary treatment process by either physical, chemical or biological means to remove nutrients that cause eutrophication. Tertiary treated waste may be treated to a level at which it can be re-used by humans. This is often described as a 'zero discharge' option.

In the Albany case, the most important process sustainability principles are likely to be triple-bottom-line integration, and accountability, transparency and engagement. Recent literature has shown that careful design of public participation exercises is an integral part of developing accountability and transparency (Webler *et al.* 2001).

The interest groups

Water Authority

Involved Water Authority staff are almost entirely professional engineers. The project is being run from the Authority's Albany office, with input from Perth where required.

Salmon fishers

Family groups of professional salmon fishers have earned their living from beach netting for generations. They are concerned about the effects ocean outfalls could have on the migratory patterns of salmon along the coast. Fisheries scientists claim that there is evidence to show that salmon move away from the coast to deeper water when they encounter fresh water.

Abalone divers

A select group of abalone divers operate, as individuals, all around the Great Southern coastline. Abalone are very sensitive to environmental contamination. Abalone licences trade for approximately $1 million, but they are not harvested near Sand Patch, so divers have no concerns about this option.

Conservation groups

There are a number of very active conservation groups in the Albany area, all of which are entirely opposed to the Water Authority's ocean outfall plans. Members cover the socioeconomic spectrum, but the most active players tend to be teachers, public servants and retired people.

Environmental Protection Authority

The EPA has had a long-term interest in the environmental condition of the Albany Harbours, and its Marine Investigations Branch undertook detailed studies of water quality throughout most of the 1980s. It is not concerned about which option is chosen, as long as the King Point outfall and treatment plant is closed and a new system is operating by the end of 1994. However, the EPA will impose discharge standards on the Water Authority when it licenses the new treatment and disposal system.

Environmental Consulting company

Early in 1991 the Water Authority chose an environmental consulting company to undertake the required environmental impact assessment (EIA) on the chosen option. The company chosen by the Water Authority has offered a team made up of two marine biologists, a waste-water treatment engineer, and a social scientist. The company has a client relationship with the Water Authority and must advocate for them. It also, however, has a professional responsibility to undertake an objective environmental assessment.

Surfers

Surfers use the area close to the proposed Sand Patch outfall extensively. This beach is remote, although it can be reached by four-wheel-drive tracks west of Torndirrup National Park. Some surfers are vocal opponents of the Sand Patch option. No surfing is to be had at Ledge Beach or Nanarup Beach.

Town of Albany

This local government is responsible for the urban areas of Albany. The Town of Albany's jurisdiction does not extend to any of the ocean outfall areas but it may encompass land disposal areas if the latter is close to the centre of Albany. Ninety per cent of waste-water is generated by people living and/or working in the Town of Albany.

Shire of Albany

This local government is responsible for the rural areas surrounding the Albany townsite. The Shire of Albany has jurisdiction over all of the ocean outfall sites. It has a history of conflict with the Town of Albany.

Department of Conservation and Land Management

The Department of Conservation and Land Management (CALM) is responsible for the management of a National Park close to the Ledge Bay area. The proposed pipeline to Ledge Beach would cross the National Park. This area also contains Declared Rare Flora.

Timing

Participants in this case-study exercise should consider that the period during which they are attempting to resolve this conflict is early 1992. The Water Authority was required to act by the end of 1994 to meet EPA requirements. When an option is finally chosen, it must be presented to the EPA in the form of an EIA. The *Environmental Protection Act 1986* (WA) requires project proponents, in this case the Water Authority, to present the EIA for formal public review.

In total it would take the Water Authority approximately eight months to complete the EIA on the chosen option. Allowing for a public review period of up to ten weeks, an assessment by the EPA and issue of EPA recommendations, and the assignment of Ministerial Conditions, the project could quite easily take 12–18 months before it obtains formal go-ahead.

Conclusions

This case study contains many useful insights into the sustainability conundrum. It requires students to analyse a public decision-making process that clearly went wrong in the early stages. When taken to its conclusion as an interactive public meeting, students often arrive at a different technical solution to the Albany waste-water management problem than that which was originally proposed by the Water Authority. This also tends to make students aware that the original decision-making process chosen by the Authority was not, in itself, sustainable. Another outcome of the case, therefore, can be discussion about the nature of process sustainability.

Appendices

The following should also be studied: A. Further details on marine and land disposal options; B. Environmental values and impacts associated with the three marine disposal options; C. Capital costs and net present values of options using a 4% discount rate; D. Glossary; and E. Ideas for instructors.

A. Further details of marine disposal options

Sand Patch

- The discharge of secondary treated effluent, subsequently disinfected (bacterial levels effectively zero), or tertiary treated effluent from a shoreline outfall 700 m to the west of the main Sand Patch Road

- Effluent would be conveyed to the site from waste-water treatment plant No. 2 or No. 5 or both

- Final pipeline route would follow the present prison pipeline, descend to the beach and cross to the beginning of the surf zone

Ledge Bay

- The discharge of secondary or tertiary treated effluent through an outfall and diffuser into Ledge Bay

- The pipeline route would convey effluent from waste-water treatment plant No. 2 or No. 5 or both, across Oyster Harbour and around the landward side of Mt Martin to Ledge Beach

- The outfall would be constructed through the dunes near the end of Ledge Beach Road and continue seaward for 1,100 m

Nanarup

- The discharge of secondary or tertiary treated effluent through an outfall and diffuser off Nanarup Beach

- The pipeline route would convey effluent from waste-water treatment plant No. 2 or No. 5 or both, follow the road reserve over bridges crossing the King and Kalgan rivers across the top of Oyster Harbour, and then along the Nanarup Road

- The outfall would be constructed through dunes 100–200 m east of Taylor Inlet and extend offshore for a distance of 1,300 m

B. Environmental values and impacts associated with the three marine disposal options

a. Frequency/intensity of human uses

	Site		
	Sand Patch	*Ledge Bay*	*Nanarup*
Swimming	–	++	++
Surfing	++	–	+
Diving	–	+	++
Aesthetic enjoyment	++	++	+++
Beach walking	+	–	++
Hang gliding	+	–	–
Recreational fishing			
Shore-based	++	–	++
Boat-based	–	++	–
Abalone	–	–	+
Spear fishing	–	+	++
Commercial fishing			
Salmon	–	–	+++
Pilchard	+++	+	++
Abalone	–	–	–
Other	–	–	–

–   Minor or non-existent
+   Moderate/low
++ Considerable in parts
+++ Heavy

b. Beneficial uses

	Site		
	Sand Patch	*Ledge Bay*	*Nanarup*
Direct contact	+	++	++
Harvesting aquatic life (non-mollusc)	+	++	++
Harvesting molluscs	n.a.	n.a.	+
Passage of fish	++	n.a.	++
Preservation of ecosystem	++	++	++

n.a. Not applicable
+   Applies to parts of the area only
++ Applies to whole area

c. Environmental impacts predicted to arise from disposal of secondary treated waste-water at marine sites evaluated

	Site		
	Sand Patch	Ledge Bay	Nanarup
Negative impacts on			
Aesthetic enjoyment	+	–	–
Fishing	?	–	?
Algal growth	–/?	++*	–
Seagrass	–	+*	–
Public health	–	–	–
Adjacent coast	?	+	–

—   No impact
+   Minor impact
++ Major impact
?   Impact may occur—uncertain
*   Mitigated by use of superior tertiary treated effluent

C. Capital costs and net present values (NPV) of options using a 4% discount rate ($ millions)

Disposal options	Initial capital*	Total capital**	NPV capital and operating	NPV income of farm	NPV
Sand Patch					
Secondary	14.17	37.32	36.69		36.69
Tertiary	14.77	40.04	45.23		45.23
Zero-discharge tertiary	15.97	44.84	48.90		48.90
Ledge Beach					
Secondary	20.42	50.45	44.93		44.93
Nanarup					
Secondary	24.88	61.63	53.12		53.12
Land disposal	15.79	41.28	41.28	3.56	37.72

Notes
● Effluent would be disinfected for all Sand Patch options.
● Tertiary treatment reduces nitrogen to 10 mg/l and phosphorus to 3 mg/l.
● Zero discharge tertiary reduces both nitrogen and phosphorus to 0.5 mg/l.

*   Expenditure in first 3 years
** Expenditure until 2015

D. Glossary

Activated sludge process	The entire process by which aerated bacteria from recycled sludge convert the organic component of waste-water into bacterial mass and stabilised compounds such as nitrates, sulphates and carbon dioxide. This process is at the heart of the secondary treatment step
Aerated ponds	A secondary, biological waste-water treatment technique, intermediate between waste stabilisation lagoons and activated sludge processes. They entail artificial means of aeration, but do not include a sludge recycling step
Aeration	The first stage in secondary treatment and a component of the activated sludge process. Aeration, either by agitation or trickling filtration, brings the organic matter into contact with sludge which is heavily laden with bacteria
BOD	Biological oxygen demand—an indication of the amount of oxygen needed to oxidise the organic matter in a water sample by biological means
Land treatment (or disposal) system	A system that utilises the filtering, bacterial and adsorbent properties of soil and vegetation to remove impurities from waste-water
Primary treatment	The first step in the treatment of waste-water. It includes screening sedimentation, and sometimes—although rarely—chemical precipitation
Secondary treatment	A biological treatment process, subsequent to primary treatment. Secondary treatment is very similar in concept to the processes of decomposition in nature. Organic wastes are consumed by bacteria under controlled conditions so that most of the BOD is removed in the treatment process. Secondary treatment can utilise waste stabilisation ponds, land treatment, aerated lagoons, or the activated sludge process
Sewage	Raw, untreated effluent/waste-water
Sewerage	The physical system (that is, pipes, pumps, etc.) by which sewage is transported
Tertiary or advanced treatment	Not clearly defined but generally relating to the further treatment of secondary effluent by either physical, chemical or biological means to remove nutrients that cause eutrophication. Physical techniques include filtration, distillation and reverse osmosis; chemical processes include electrodialysis, precipitation, carbon absorption, ammonia stripping and ion exchange; biological processes include the harvesting of algae grown on nutrients and bacterial nitrification and denitrification. Tertiary treated waste may sometimes be treated to a level at which it can be re-used by humans. This is often described as a 'zero discharge' option
Waste stabilisation ponds (lagoons)	A secondary, biological waste-water treatment technique intermediate between land treatment and other more controlled forms of biological treatment such as aerated lagoons or activated sludge. Lagoons require little experience to operate but demand large areas of land and provide little control over process effectiveness

E. Ideas for instructors

a. Structuring the role-play

In our experience, the most effective way to run this case as an interactive exercise is to organise it as a public meeting.

If this approach is taken, a Water Authority engineer can run the public meeting, presenting the Sand Patch cliff outfall as the Authority's chosen option. This will lead to arguments inside the meeting format. The roles outlined above under the heading 'The interest groups' can be taken by one or two students. If there are lots of students, those not assigned to a role can play general 'members of the public'.

We recommend that the instructor take the role of the Water Authority engineer running the public meeting (and thus the development of the case itself).

b. Timing

Timing can be variable, depending on how long students are given to prepare their roles. In our experience, the minimum amount of time it takes to run the case is 50 minutes. This can be easily expanded to 120 minutes.

c. Leading to an outcome

While this role-playing case exercise can take on a life of its own, our experience has indicated that there are advantages for the instructor in attempting to lead towards a conclusion.

Our favoured approach is to have the interest groups argue their positions for 30 minutes or so. This tends to accentuate the differences, and lead to an unresolved, 'locked' outcome.

At this point it may be best to halt the role-playing exercise, and begin to analyse the case outcomes, and the public meeting process.

d. Analysing the outcomes

Perhaps the first thing to do is to ask the students how the Water Authority engineer could extract him/herself from this difficult situation.

Sometimes the role-playing participants realise that land disposal may be a good solution that was not properly investigated in the Water Authority's earlier work. If this outcome is reached, then the Water Authority engineer can suggest the commissioning of a consultant to further investigate feasibility and to suggest sites. This is, in fact, what happened in practice. This is one way of addressing the 'accountability' process sustainability principle introduced earlier.

If the possibility of land disposal does not present itself in the public meeting, then the Water Authority engineer does not have too many options for saving face.

In reality, the Water Authority did employ consultants to further investigate land disposal. The consultants produced a study that identified a small number of sites that were not waterlogged, and had the correct combination of soils for binding phosphorus (nitrogen is bound by plants). This options study became the basis for a full environmental impact assessment document, which was presented to the EPA for approval.

Approval was granted, and Albany waste-water is now secondary-treated and used to grow hardwood trees for woodchip export.

This was a significant cultural change for the Water Authority. To its credit, it made the jump from a 'hard' engineering organisation that traditionally saw ocean outfall as the only way to deal with this kind of problem, to an organisation that could accept land disposal and tree farming as part of its remit. The Authority also learned that sustainable outcomes from controversial development proposals can be achieved only by addressing decision-making process issues.

e. Questions for students

The case offers a number of possible learning objectives for instructors. We have used the case to amplify issues such as: conflict resolution theory and practice, the structure of the environmental regulatory framework, and the importance of 'process' in moving towards sustainability.

Questions that might be put to students when analysing the case include:

- What are the main characteristics of this conflict? Are they mostly substantive, procedural or psychological?

- What strategies could be used to slow down this conflict?

- How could the Water Authority have done a better job in the first place? Here the issue of structuring the decision-making process is important. The Water Authority could have arguably done a better job earlier on by allowing public input into the original options choice exercise. This would have met the 'accountability/transparency/engagement' process sustainability principle, and the 'triple bottom line' principle. Techniques such as multi-criteria analysis could have been used to structure this open options choice exercise (Annandale and Lantzke 2000)

f. Providing materials to students

Students should be provided with all materials up to the end of the section above titled 'The interest groups', prior to the case study. They need to have enough time before the running of the case to properly understand their roles.

As mentioned above, the 'additional financial resources' should be given out at some point during the public meeting, preferably when the Water Authority engineer is trying to make a point about costs.

The newspaper cuttings should be shown only at the end of the analysis.

g. Additional resources

Additional financial information is attached at the end of this section. A set of two tables converts the financial information given in Section C above to rate increases. You should distribute this information whenever you think that it might assist discussions, but it should probably not be distributed until the public meeting is well under way. The Water Authority engineer can use the tables to argue against tertiary treatment from a cost point of view. It could also be suggested that individual ratepayers would have to pay additional costs, if Albany decides to go beyond the Sand Patch secondary treatment option.

Additional information by way of newspaper clippings is also attached at the end of this section. There are two clippings:

- Effluent, trees a winning combination (see Fig. 15.2)

- How the Albany community helped us clean up at the environmental awards (see Fig. 15.3)

- Because these clippings point to the benefits of the land disposal option, they should be kept from students until the end of the case exercise.

Effluent, trees a winning combination

Jumping for joy: Treatment plant operator Brian Green leaps over one of the thousands of blue gums that are fed on effluent. PICTURE: JOHN EVANS

BY CARMELO AMALFI

A PRIZE-WINNING "living laboratory" of 500,000 fast-growing trees is expected to handle Albany's effluent disposal demands until at least 2020.

The trickle of treated effluent on experimental eucalypts earned WA's Water Corporation the Australian Water and Wastewater Association's Environment Merit Award, announced in Melbourne this week.

The award recognises the WA authority's innovative solution to an environmental problem facing the world — how to reuse wastewater.

The first 300ha of mainly Tasmanian blue gums near Albany airport, 10km north of the town, will be ready for harvest and sold by 2002 — having fed on 3000 kilolitres of treated effluent pumped more than 10km each day from the King Point treatment plant.

The project was developed in 1993 to reduce the flow of nutrients into Albany's Princess Royal Harbour and King George Sound. It was set up after intense community debate over water pollution.

Managing director Jim Gill said he was delighted that the corporation had been recognised nationally for its environmental leadership in waste disposal technologies.

"That is exactly what we wanted to be — responsible environmental managers," he said.

"The award gives us a position of leadership in Australia and opens up new doors internationally and locally in terms of intellectual property and consulting opportunities."

Mr Gill said 12 other sites, mainly in the South-West, were being considered in plans to expand the project in WA, starting with Manjimup.

Water Resources Minister Kim Hames, who described the scheme as a pacesetter in wastewater reuse, said the award was a fitting tribute to the community whose support was crucial.

Bob Silifant, Great Southern region manager of field support services, said Albany's $18 million tree farm project also had become a popular tourist and educational attraction for Australian and overseas visitors.

The computer-controlled liquid fertiliser farm would create jobs and revenue from a woodchip industry based on quality paper pulp from trees that could grow to 15m within seven years.

Barry Sanders, general manager of bulk water and wastewater, said the biennial award was the town's second.

The first was scooped up five years ago by engineering consultant Sinclair Knight Mertz, which developed a way to improve the quality of CSBP effluent entering local waters.

About 250 million litres of wastewater are collected each day in WA. Only five million litres are recycled.

FIGURE 15.2

Tasmanian Blue Gums at the Albany Tree Farm.

How

the Albany community helped us clean up at the environmental awards.

Thirteen kilometres north of Albany you'll come across a rather unique, 400 hectare farm.

Its produce is trees - heavy-drinking Tasmanian blue gums.

What makes this farm unique is that the trees are thriving on what was once a threat to the local environment - wastewater being discharged into Albany's harbour.

Thanks to the wholehearted support of the local community

that wastewater is now pumped from the Albany Wastewater Treatment Plant to the tree farm where it trickle irrigates some 500,000 Tasmanian Blue Gums.

The Albany Tree Farm is an innovative solution to an environmental problem, and a shining example of the Water Corporaton and the community working together to help protect our fragile environment.

WATER
CORPORATION

FIGURE 15.3

Additional financial information

1. **Charge increases necessary to recover capital and operating costs (assuming a present residential rate of 6.23¢/$ gross rental value [GRV] and non-residential rate of 6.98¢/$GRV).**

Disposal options	Residential (¢/$GRV) increase	Non-residential (¢/$GRV) increase
Sand Patch		
Secondary	2.95	3.31
Tertiary	3.69	4.13
Zero-discharge tertiary	4.00	4.49
Ledge Beach		
Secondary	3.66	4.10
Nanarup		
Secondary	4.37	4.89
Land disposal	3.04	3.41

2. Examples of range of increase in annual rates for typical properties ($)

Property	Rental value	Increase residential		Increase non-residential	
	(RV)	R_1	R_2	NR_1	NR_2
Houses					
Chester Pass Rd	3,588	86	156		
Parade St	4,264	103	186		
Spencer Park	5,408	130	236		
Serpentine Road	5,928	143	259		
Commercial properties					
Shop York St	29,640			800	1,449

Notes
R_1 Rate increase of 2.41¢/$RV land disposal/Sand Patch secondary
R_2 Rate increase of 4.37¢/$RV Nanarup
NR_1 Rate increase of 2.70¢/$RV land disposal/Sand Patch secondary
NR_1 Rate increase of 4.89¢/$RV Nanarup

References

Annandale, D., and R. Lantzke (2000) *Making Good Decisions: The Role of Decision-Aiding Techniques in Waste Facility Siting* (Perth, Australia: Murdoch University).

EPA (Environmental Protection Authority) (1990) *Albany Harbours Environmental Study 1988–1989* (A Report to the Environmental Protection Authority from the Technical Advisory Group; Bulletin 412; Perth, Australia: Environmental Protection Authority).

Government of Western Australia (2002) *Focus on the Future. The Western Australian State Sustainability Strategy: Consultation Draft* (Perth, Australia: Department of the Premier and Cabinet).

Webler, T., S. Tuler and R. Krueger (2001) 'What is a Good Public Participation Process? Five Perspectives from the Public', *Environmental Management* 27.3: 435-50.

About the contributors

David Annandale is Senior Technical Advisor to the National Environment Commission, Kingdom of Bhutan. At the time of writing, he was Senior Lecturer in Environmental Assessment and Policy, School of Environmental Science, Murdoch University, Perth, Australia.

d.annandale@murdoch.edu.au

Shirley Eber has been teaching a variety of tourism-related subjects in universities in the UK and abroad for over 15 years. Undergraduate and postgraduate courses include sustainable tourism management, sustainable business, tourism and development, and the anthropology of tourism. She has worked as a tourism education consultant, and is a published author, and editor and translator. Her undergraduate studies in Arabic and French, followed by a master's degree in the Anthropology of Tourism, consolidated her extensive experience of travel and interest in exploring the wider implications of the tourism business on both a personal level and for society at large.

shirley@eber.fsworld.co.uk

Jill Engel-Cox, PhD, is a senior research scientist at Battelle. She has taught pollution prevention courses at Johns Hopkins University and Washington State University-Tri-Cities, including student-conducted pollution prevention assessments at local small businesses. For the US Environmental Protection Agency, she restructured and finalised the training materials for the 'International Principles of Pollution Prevention and Cleaner Production' training course, teaching it in both Washington, DC, and Shanghai, China. She has facilitated numerous pollution prevention and environmental assessment training sessions at industrial sites, universities and national laboratories.

engelcoxj@battelle.org

David K. Foot is Professor of Economics at the University of Toronto in Canada. Since his doctorate in economics from Harvard University, Professor Foot's research and teaching interests have increasingly focused on economic–demographic interactions, particularly the economic and policy implications of population ageing. He is author of the Canadian best-selling books under the *Boom, Bust and Echo* title. Professor Foot is a recipient of the national 3M Award for Teaching Excellence and is a two-time winner of the undergraduate teaching award at the University of Toronto. He is a much-sought-after speaker to businesses, associations and governments on the implications of demographic change, both nationally and globally.

foot@chass.utoronto.ca

Kim M. Fowler is a senior research engineer at the Pacific Northwest National Laboratory. She teaches pollution prevention and sustainability-related courses at Washington State University-Tri-Cities and provides pollution prevention and sustainable design training to research and operations personnel for government and private industry clients. Outside of work she is focused on enhancing Washington State K-12 science education by adding interactive, systems-focused sustainable design activities to the curriculum and through increased scientific and business community involvement in the science programmes.

kim.fowler@pnl.gov

Chris Galea is a father, educator, outdoor enthusiast, builder, sailor and entrepreneur. He currently manages his own hospitality business and also teaches at the Gerald Schwartz School of Business at St Francis Xavier University in Antigonish, Nova Scotia, Canada. Chris was part of the founding faculty of the Sustainable Enterprise Academy at the Schulich School of Business at York University in Toronto. Much of his doctoral and research work is in the area of management learning as it relates to sustainability. Chris lives by the ocean that he loves, surrounded by land he cherishes and people he deeply cares for.

cgalea@stfx.ca

Darcy Hitchcock and **Marsha Willard** are founders of AXIS Performance Advisors, a consulting firm that has been in business since 1990. They apply their management consulting, training and facilitation skills to help organisations find responsible solutions that meet all stakeholder needs: for owners, customers, employees, the community and the environment. They have co-authored six popular business books on such topics as teamwork, trust, work redesign and quality. Their latest book, *The Business Guide to Sustainability* (Earthscan, 2006), shows organisations how to integrate sustainability into their business practices. They are recognised experts in the implementation of sustainability inside organisations. Their Sustainability Series™ booklets show organisations how to simultaneously improve their financial, social and environmental performance. They designed and facilitated a professional certificate programme on Implementing Sustainability for Portland State University. Darcy teaches and does presentations for the Oregon Natural Step Network. Marsha facilitates multi-stakeholder consensus processes and teaches for several universities. (See www.axisperformance.com.)

darcy@axisperformance.com

marsha@axisperformance.com

Kumba Jallow is a principal lecturer at Leicester Business School, De Montfort University, UK. Her research interests include corporate social responsibility and accounting for sustainability. This is also reflected in her teaching, which is to both undergraduates and postgraduates. She has published widely and is the Editor of *The Journal of Applied Accounting Research*.

lhacc@dmu.ac.uk

John MacKinnon is currently a Manager with Deloitte Touche Canada's Consulting Practice. He has over seven years of strategies and operations experience. Areas of expertise include project management, process redesign, channel strategies, content management, service transformation and systems implementation. John is a Certified Management Consultant and Certified Project Management Professional. He holds a Master of Business Administration from Queen's University and a Bachelor of Information Systems from St Francis Xavier University, Canada.

Angus Morrison-Saunders is Senior Lecturer in Environmental Impact Assessment, School of Environmental Science, Murdoch University, Perth, Australia.

a.morrison-saunders@murdoch.edu.au

Christoph Otte is based at the Carl von Ossietzky University Oldenburg in the Department of Economics and Law.

christoph.otte@web.de

Joshua Skov has an MA in Economics from the University of California, Berkeley, an MA from the University of Washington and a BA from Yale University. At Good Company, a research and consulting firm that helps clients measure, manage and market their social and environmental performance, Joshua has spearheaded the development of indicators and assessment tools for a variety of organisations. Joshua appears regularly at conferences on campus sustainability and other sustainability topics. In addition, Joshua is an adjunct instructor in the Department of Planning, Public Policy and Management at the University of Oregon. Joshua also serves as co-chair of the Life Cycle Analysis committee for the Sustainable Products Purchasers Coalition.

joshua.skov@goodcompany.com

Peter Stanwick is an Associate Professor in the Department of Management at Auburn University in Auburn, Alabama. He received his undergraduate degree from the University of Western Ontario, his Master of Business Administration degree from the University of Washington, and his doctoral degree from the Florida State University. At Auburn University he teaches strategic management at undergraduate and graduate levels and organisational theory at the doctoral level. His research focuses on environmental issues, corporate governance and ethics as they relate to business.

stanwpa@business.auburn.edu

Sarah Stanwick is an Associate Professor in the School of Accountancy at Auburn University in Auburn, Alabama. She received her undergraduate degree from the University of North Carolina at Greensboro, her Master of Accountancy degree from the University of North Carolina at Chapel Hill, and her doctoral degree from the Florida State University. At Auburn University she teaches financial and cost accounting. Her research explores social issues, including corporate governance and environmental issues, and their relationship to accounting. She is a Certified Public Accountant.

stanwsd@auburn.edu

Susan Svoboda holds an MBA from the University of Michigan. She is the Managing Director of Realia Group, a strategy-training group and co-developer of Transformation: The Business Strategy Laboratory. She has led training workshops for managers from business organisations and at universities. She is also an adjunct faculty member at Georgetown University at the School of Foreign Service and the Business School and an instructor for the American Society of Quality. Prior work experience includes consulting in the field of organisational effectiveness and performance benchmarking, and purchasing in the retail sector.

Susan@realiagroup.com

Dipl. Oek. **Anke Truscheit** has been at the Carl von Ossietzky University Oldenburg since October 1998 where she is Co-ordinator of the Ecological Economics course in the Department of Economics and Law.

anke.truscheit@uni-oldenburg.de

Penny Walker is an independent consultant specialising in helping people in business learn about sustainable development, and put that learning into action. Formerly a campaigner with Friends of the Earth, she lives in London with her family.

penny.walker@btclick.com

John Whalen is a partner with Sustainable Value Partners, Inc., a consulting and education firm helping corporations create value for shareholders and stakeholders. John has over 20 years' experience in organisational innovation and whole-systems change and has used experiential learning in support of organisational and leadership development for many years. John's current focus is helping corporations create competitive advantage by integrating sustainability into their core business strategy.

john@sustainablevaluepartners.com

Monika Winn is Associate Professor of Business Strategy and Sustainability. Her research focuses on organisational and institutional change related to the challenges of sustainability (including climate change impacts on strategy, marketing for the poorest of the poor, stakeholder conflicts, and social to strategic issue transformation (salmon farming). Monika has published in *Organization Studies*, *Academy of Management Review*, *Business and Society* and *The Journal of Business Venturing*. She co-founded and chaired the Academy of Management's 'Organisations and the Natural Environment' group.

MIWinn@UVic.CA

Charlene Zietsma is an assistant professor of Strategy at the Ivey Business School, University of Western Ontario. She completed her PhD at the University of British Columbia. Charlene's research interests focus on processes of emergence and change, especially in the context of corporate conflicts with social and environmental stakeholders. She has studied processes of institutional change, organisational learning, dynamic capability emergence, and opportunity recognition. She has served on a number of boards and committees for non-profit, professional and government organisations.

czietsma@ivey.uwo.ca

Index